T0265589

BLOOD ON THE COAL

BOOKS BY KEN CUTHBERTSON

AS AUTHOR
Inside: The Biography of John Gunther
Nobody Said Not to Go: The Life, Loves, and
Adventures of Emily Hahn
A Complex Fate: William L. Shirer and the American Century
The Ring of Truth: The Memoirs of
the Hon. Henry E. MacFutter (fiction)
The Halifax Explosion: Canada's Worst Disaster
1945: The Year That Made Modern Canada
When the Ponies Ran: The Untold Story of Kingston's Minor-
League Professional Baseball Team, 1946–51

AS EDITOR
Congo Solo: Misadventures Two Degrees North by Emily Hahn
Queen's Goes to War: "The Best and the Worst of Times"

KEN CUTHBERTSON
BLOOD ON THE COAL

THE TRUE STORY OF
THE GREAT SPRINGHILL
MINE DISASTER

HarperCollins*Publishers*Ltd

Published by HarperCollins Publishers Ltd

First edition

HarperCollins Publishers Ltd
Bay Adelaide Centre, East Tower
22 Adelaide Street West, 41st Floor
Toronto, Ontario, Canada
M5H 4E3

www.harpercollins.ca

Photo on page iii: View from the pithead (Clara Thomas Archives and
Special Collections, York University)
Picture of miner's helmet by Shutterstock
Map by Larry Harris

Library and Archives Canada Cataloguing in Publication

Title: Blood on the coal : the true story of the great Springhill mine disaster /
Ken Cuthbertson. | Names: Cuthbertson, Ken, author.
Description: Includes bibliographical references and index.
Identifiers: Canadiana (print) 20230438059 | Canadiana (ebook) 20230438083
ISBN 9781443467919 (hardcover) | ISBN 9781443467926 (ebook)
Subjects: LCSH: Coal mine accidents—Nova Scotia—Springhill.
LCSH: Miners—Nova Scotia—Springhill.
Classification: LCC TN311 .C88 2023 | DDC 971.6/11—dc23

Printed and bound in the United States of America
23 24 25 26 27 LBC 5 4 3 2 1

To
Harold Brine, Maurice Ruddick,
Raymond "Tommie" Tabor, Dr. Arnold Burden,
and the people of Springhill, Nova Scotia

and

Britton C. Smith,
a good friend who passed much too soon

The extraction of oil, coal and minerals brought,
and still brings, a cost to the environment.
—BONO

The only truly dead are those
who have been forgotten.
—JEWISH PROVERB

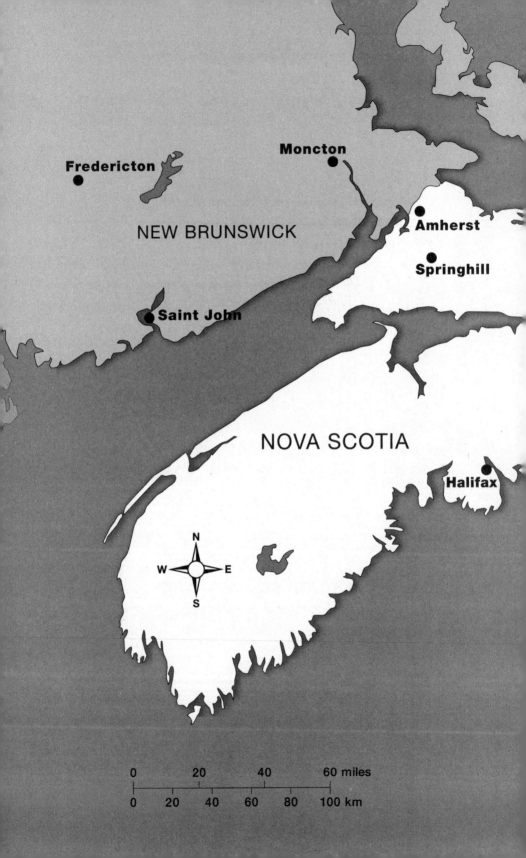

PRINCE EDWARD
ISLAND

Glace
Bay

Sydney

New Glasgow

THE SPRINGHILL
MINE DISASTER

OCTOBER 23, 1958

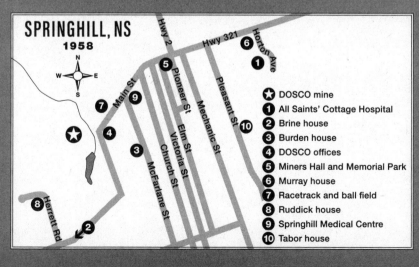

SPRINGHILL, NS
1958

Hwy 2

Hwy 321

Horton Ave

Main St

Pioneer St

Elm St

Victoria St

Church St

McFarlane St

Mechanic St

Pleasant St

Herrett Rd

★ DOSCO mine
1 All Saints' Cottage Hospital
2 Brine house
3 Burden house
4 DOSCO offices
5 Miners Hall and Memorial Park
6 Murray house
7 Racetrack and ball field
8 Ruddick house
9 Springhill Medical Centre
10 Tabor house

CONTENTS

FOREWORD
BY ANNE MURRAY

I CHERISH MY MEMORIES OF LIFE IN SPRINGHILL IN THE 1950s. I came from a big, happy family of six with loving parents. My dad was one of the town doctors, while my mom had worked as a nurse until she had children.

We lived right at the top of what's called Monument Hill. Growing up there, in a home with five brothers and a dad who loved the outdoors, it isn't surprising that I was a tomboy and was forever active. Dad started throwing balls back and forth with us all at early ages. I could throw every kind of ball there was.

There was always something to do in Springhill. We had a movie theatre and a rink where I skated several days a week. I enjoyed school, had lots of friends, played softball, and, of course, I loved to sing. Despite the happiness in my life, I knew there was a dark side to life in a coal mining town.

Dad never talked about his patients, but I knew he often operated on miners who'd been injured on the job. Some of them were the fathers, brothers, uncles, or other relatives of kids I went to school with. My own maternal grandfather—Arthur Burke—had been a coal miner, and so I knew that miners flirted with death every time they went to work. That lesson was driven home to me in 1956, when I was eleven. That year, a huge explosion at the No. 4 mine killed thirty-nine miners. Then in 1957, around Christmas time, a fire burned much of the downtown. It almost seemed as if Springhill were cursed. We prayed that nothing else bad could happen to the town, but unfortunately, something did.

On October 23, 1958, the big Bump occurred. Seventy-five men died, and the No. 2 mine—the last of the town's mines that was still open—was so badly damaged that it closed forever. Life in Springhill was never the same again.

I still get teary, and it chokes me up whenever I think back to that time or when I talk about it. I have vivid memories of standing around at the pithead for hours with girlfriends whose fathers were trapped in the mine. We waited for news all day in the cold and the rain. We'd go home to warm up, dry out, and have something to eat, and then we'd go right back to join the vigil. We rejoiced when we heard a miner was saved, and we felt devastated when word came that another body had been found. Many of my friends lost their fathers in that disaster, and I remember feeling sad beyond words.

In the years when the Springhill mine was a going concern, the miners made a decent living, and Springhill was a good place to live. All that changed after the Bump. There was a pall over the town.

I was fortunate. Life didn't seem to change all that much for me or my family after the disaster. Many families left town or moved away. We stayed. My dad continued his medical practice, and he and my mom remained vitally involved in the community. My brothers and I grew up, went off to university, and eventually moved on, but Springhill was still "home" to us.

It took a few years, but people there got on with life and began to hold up their heads again. Springhillers are resilient. They've always taken pride in their community. I know I'm still proud to call myself a Springhiller. I always will be.

2

This year marks the sixty-fifth anniversary of the Springhill mine disaster. The world has changed a lot since 1958, but in some ways, not all that much has changed. All of the coal mines in Nova Scotia had been closed for years until recently, when one reopened in Donkin, Cape Breton. New mines are being opened elsewhere in the world too. People are still burning coal in many countries, and miners are still dying in cave-ins, bumps, and accidents.

I think there are lessons to be learned from what happened in Springhill, but it does seem we've failed to learn them. Ken Cuthbertson makes that point in the book you're about to read. I think he's right in arguing that the story of the final Springhill mine disaster is a cautionary tale that's worth remembering.

In 1958, it was my hometown that was in peril. Today, it's the world.

A DISASTER WAITING TO HAPPEN

DEATH CARES NOTHING FOR ETIQUETTE. ALL TOO OFTEN, like the unwelcome visitor that it is, death arrives unannounced and unexpectedly.

It did that on the evening of October 23, 1958. One hundred and seventy-four miners were at work deep within the Dominion Steel and Coal Corporation (DOSCO) colliery at Springhill, Nova Scotia, when the earth suddenly moved. Scores of men died instantly. Others escaped death only because they somehow scrambled to safety. Still others, two groups totalling twenty men, found themselves trapped deep underground in the eternal darkness. Many of them had suffered serious injuries; several were clinging to life by the slimmest of threads. All prayed for a miracle rescue.

"At the surface [in Springhill], people . . . felt a bump," a Nova Scotia Energy and Mines senior geologist would say many years later. "That wouldn't explain what the miners felt deep underground. It was much more violent."[1]

Shock waves from the killer tremors took just seventeen seconds to travel 110 miles through the earth to the seismograph at Dalhousie University in Halifax, and a little more than a minute to reach Ottawa, 800 miles to the west.

It has been guesstimated that the force of what locals ever after would refer to as "the Bump"—as it became known, with a capital B—was the equivalent of about a thousand tonnes of dynamite being exploded underground. That may well have been the case, for the dire consequences of the resulting upheaval stand to this day as one of Canada's worst-ever workplace disasters. Springhill, a hard-luck town if ever there was one, had a long, painful history of such suffering.

Whenever there was an emergency at DOSCO's Number 2 (No. 2) coal mine at Springhill, a whistle usually sounded. However, on this particular evening, that alarm was silent, at least initially. Not that it really mattered. The instant the townspeople heard the rumble and felt the ground shake, they knew there was trouble at the mine. Big trouble.

On the streets of town, the tremors were strong enough to make weathered wooden buildings creak and groan. Window glass shattered. Spiderweb-like cracks scored plaster walls and ceilings. Dishes chattered on kitchen shelves. Lights flickered. Televisions went dark. Some townspeople were rendered speechless; children cried out in fear. Dogs whimpered as they scrambled to find shelter.

In Miners Memorial Park on the town's Main Street, the iconic "White Miner" memorial statue quivered atop its 16-foot-tall plinth. For sixty-three years, that solitary marble figure had borne silent witness to the harsh realities of life in this hardscrabble, one-industry town. Ol' King Coal dominated every aspect of life in Springhill for nine decades, and regrettably, this monarch wasn't a benevolent ruler.

The threat of accidents and sudden death was a constant concern that haunted the men who toiled in the pitch-black depths of the town's No. 2 mine. Many people said it was the world's deepest and most dangerous coal pit. Those who did so were likely cor-

6

rect. But who really knows? About all that can be said for certain is that in October 1958, DOSCO's No. 2 Springhill colliery was a leading candidate for both those dubious distinctions.

The Springhill miners memorials occupy a prominent location on the town's Main Street. (Courtesy of Wally Hayes)

The mine's vertical depth from the pithead down to its nethermost reaches was 4,600 feet. Do the math, and you'll discover that's more than twice the height of Toronto's CN Tower. At the same time, the mine's slopes stabbed into the earth at a precipitous 30-degree angle, giving miners access to the rich seams of bituminous coal that honeycomb the substrata of Nova Scotia's craggy Cumberland County. Each day the miners went to work, they rode a trolley westward for more than 2.6 miles, down, down, down into the bowels of the mine, travelling along "roads that never saw sun nor sky."[2]

The deeper the No. 2 mine went, the more unstable and dangerous the earth around it became. Over the four decades prior to 1958, more than five hundred mini-earthquakes had shaken the mine—"bumps," that's what the miners called them.

It was in hopes of reducing the frequency of those bumps and increasing the productivity of the money-losing mine—which reportedly was in the red by more than $700,000 so far that year[3]—that Louis Frost, the coal company's chief mining engineer, had expedited changes in the way coal was being extracted. Although his decision to do so was based on the pressing need to boost production and reduce the frequency of bumps, it was controversial nonetheless. The men who risked their lives every day in the No. 2 mine insisted that while Frost's strategy might reduce the number of bumps, it would also increase the likelihood of a jolt of epic intensity. One veteran miner voiced the dread that he and many of his co-workers felt when he protested, "The college boys didn't believe us when we told them that if they continued doing things that way, they were going to kill us all."[4]

Many of the miners feared that the No. 2 mine was the proverbial "accident waiting to happen." And when it did, the consequences would be catastrophic. Hindsight tells us they had good reason to be afraid.

The seismic upheaval that occurred on the evening of October 23, 1958, was the worst of its kind ever to rock a North American mine. It destroyed the No. 2 Springhill colliery and claimed the lives of seventy-five miners. Scores of other men suffered grievous injuries; nine hundred jobs disappeared in an instant, and countless lives were ruined. The Bump ripped the soul out of the town of Springhill.

For almost two weeks, the eyes of the world focused on this hard-luck coal mining town in rural Nova Scotia. On-the-spot Canadian Broadcasting Corporation (CBC) coverage of the disaster mesmerized television viewers across North America and around the world. This was the first time that people in their living rooms could view news reports being beamed live from the scene of an unfolding disaster. Meanwhile, a vast audience also tuned in to radio news reports and read the newspaper stories chronicling the tireless efforts of rescuers who were racing to save the lives of men entombed deep underground. "Canada watched these heroic efforts," a *Globe and Mail* editorial stated. "With all its heart, Canada hopes they will succeed."[5]

In this resource-rich nation, it almost went without saying that coal mining was a hazardous occupation; Springhillers had ample reason to know this better than most. By 1958, their town had already endured two devastating mine disasters—the one in 1891 had claimed 125 lives, and the other in 1956 had killed 39 men. The names of the 164 miners who perished in those calamities are inscribed on the marble monuments in the memorial park on Main Street, along with the names of almost two hundred other men who'd died in the one-off fatal incidents that had plagued the town's mines over the years, especially "old No. 2." Many people suspected the mine was star-crossed. "Remember Springhill's big producer, / her histories of crimson hue," a poem by veteran coal miner Maurice Ruddick declaimed. "The biggest tale the old miners tell, / is the curse of the old No. 2."[6]

Those who place their faith in science tend not to believe in curses. Officials from the provincial Department of Mines and DOSCO technicians were among them. For years, they recorded the dates, times, locations, and intensities of the hundreds of bumps that rattled the mine. By studying this seismological data, they hoped to discover the elusive truths that would enable them to predict where and when the next bump would happen, and hopefully reduce the incidence of bumps or at least mitigate the damage and harm such tremors caused. However, the compilation of this mountain of information was all for naught. As computer guru Steve Jobs noted in an oft-quoted 2005 commencement address he gave at Stanford University, "You can't connect the dots looking forward; you can only connect them looking backwards."[7]

Only with technical know-how and computers that in the

9

1950s were still years in the future might anyone have been able to decipher what the numbers were saying. At the time, it was even less possible than it is today to predict when a killer bump, "the Big One," would occur.

The first hint of the disaster that loomed at the Springhill mine—"the dots," if you will—appeared a few minutes before seven o'clock on the evening of October 23. That's when the earth quivered ever so slightly. Miner Raymond "Tommie" Tabor, one of a crew of about fifty men who were working at the No. 2 mine's lowest level, was chipping away at the eight-foot-high coal seam when he felt the earth heave. The instant he did so, he stopped cold and tensed up. From somewhere out there in the subterranean darkness came a low, ominous rumble. It sounded like the growling of a giant, angry beast. This aspect of the mine's soundtrack had become hauntingly familiar to Tabor. He'd worked in the mine for almost half of his thirty-eight years of life, yet his fear of bumps sometimes still kept him awake at night and gave him nightmares when he did sleep.

Whenever Tabor worked in the deepest recesses of the mine, he found it almost impossible to miss or ignore a bump, no matter how slight it was. On this night, as Tabor and his co-workers sometimes did when the earth moved, they put down their tools and waited. It usually took about twenty minutes for the mine's ventilation fans to clear the coal dust from the air. Only then, and even though he was still shaking and fearful that another bump might be coming, would Tabor resume work. After all, what choice did he have but to do so?

Like many of the men who worked in the No. 2 Springhill colliery, Tommie Tabor had a fatalistic view of what it meant to toil there. The miners accepted that it was only a matter of time until the Big One happened. And when it did, heaven help anyone who was unlucky enough to be working underground. Either you accepted the risk, or you found some other way to earn a living. "I didn't have the good sense to be afraid," one veteran miner noted. "I just knew what I had to do, and I did it."[8]

With steady jobs that paid a living wage being scarcer than palm trees in Springhill, the quintessential one-industry company town, those who earned their living digging for coal "whistled past the graveyard." They did so hoping and praying that if and when something bad did happen, they wouldn't be there for it. At seven o'clock on the evening of October 23, Tommie Tabor breathed a sigh of relief. He knew the wee bump he'd just experienced wasn't the Big One. Others also knew that.

At the 13,000-foot level of the mine, 800 feet above where Tabor was working, overman Charlie Burton had felt the same bump. He'd watched a shower of bits of thumbnail-sized coal rain down from the roof of the mine and the air grow cloudy with a fog of coal dust. Burton stopped what he was doing, calmly pulled out the notebook he kept in his pants pocket, and as he did whenever there was a bump, he jotted down the time and a few details about its severity. Burton wasn't overly concerned about a mini-bump; nor were some of the miners who were working alongside him up and down the coal face. In fact, some of them were pleased when a mini-bump occurred, for it loosened the coal and made it easier to bring down. The more coal the miners extracted on their shift, the more chance they had to earn a production bonus.

Veteran miners Harold Brine and Maurice Ruddick knew it wasn't the Big One. They'd been around long enough to understand that if you heard and felt a bump, chances were you were going to be all right. If you didn't hear or feel the earth move, it meant either the tremor was nothing to worry about or you were already dead. If not, you'd be dead before your next heartbeat. That's how quickly death could snatch a miner away. You could be alive one moment, dead the next.

The year 2023 marks the sixty-fifth anniversary of the Springhill mine disaster. The number of people who experienced the Bump of

October 23, 1958, and who remember the names and faces of its victims, inevitably grows smaller with each passing year. That's a pity, for in many ways, the story of the disaster that befell the good people of Springhill, Nova Scotia, and their town's No. 2 colliery stands as a cautionary tale that we'd do well to heed. It reminds us that if we ignore the jangling warning bells that portend a coming storm, bad things can and do happen to good people.

Being a one-industry company town, Springhill was afflicted with tunnel vision—no pun intended. All seven thousand residents relied on the immediate economic benefits that came from their dependence on coal. They did so even though in their hearts they knew that sooner or later there'd be a steep price to pay. Springhillers were damned if they did and damned if they didn't.

There's an obvious and timely lesson to be drawn from this. However, as the British author Mary Ann Evans, better known as George Eliot, once reminded us, "All meaning, we know, depends on the key of interpretation."⁹ We sometimes see only what we want to see. And contrary to the message in that folksy aphorism, what we don't see *can* hurt us.

The gates to the DOSCO coal mine were a local landmark on Main Street in Springhill. (Mary Willa Littler/ Nova Scotia Archives)

CHAPTER I

COAL, BLOODY COAL

IT WAS THE EARLY 1790S WHEN THE FIRST NON-INDIGENOUS people visited the future site of the town of Springhill, Nova Scotia. These United Empire Loyalists (UELs) stood in awe when they did so. It's easy to understand why.

The view those few early arrivals enjoyed from their lofty vantage point, more than 650 feet high, was stunning. The hilly panorama, spread out below them for as far as their eyes could see, was verdant, vast, and wondrous to behold.

In 1926, local historian Bertha Isabel Scott—herself a UEL descendant—wrote that from up here, atop one of the high points in the Cobequid Hills, "the long, unbroken horizon from West to North shows ridge upon ridge of blue hills extending to New Brunswick, and against these, the nearer forests show the changing colours of the seasons. The sunsets back of those ridges . . . are indescribably beautiful."[1] "On a clear day," as an old show tune gushes, "you can see forever." No, really—you can. It's always been true. It was true in 1790, it was true in 1926, and it remains true today.

The Cobequid Hills lie at the northernmost reaches of the Appalachians, the prehistoric spine that runs 2,000 miles along

North America's eastern seaboard, all the way from Alabama in the south to Labrador in the north. And the Cobequids are old. *Ancient* is a more accurate descriptor. The geological origins of the Appalachians date back more than 400 million years, to the Precambrian and Devonian eras. To put that into a more meaningful context, think of it this way: The dinosaurs that were digitally re-created in the *Jurassic Park* movies roamed the earth in the Mesozoic era, between 245 and 65 million years ago. The first humans didn't appear on the scene until about 2.4 million years ago.

At one time, the Cobequids were *real* mountains, lofty and towering as high as 10,000 feet. But eons of erosion and seismic upheavals of the earth's crust crushed, folded, and reshaped the Cobequids. Today, they're best described as "hills." Like the rest of the present-day province of Nova Scotia, the Cobequids were part of the supercontinent called Pangea.[2] This mother of all geological formations broke apart some 200 million years ago when a massive shifting of the earth's tectonic plates gave rise to the seven continents we know today.

The Cobequids' original human inhabitants were the Mi'kmaq, who referred to this area, part of their traditional territory, as *Kwesomalegek* (pronounced kway-soh-mal-gek); roughly translated into English, that means "hardwood point."[3] The term was apt, for the hillsides around Springhill were—as they still are—thick with sugar maples, beech, and yellow birch. Viewed from on high, this arboreal panorama is magnificent both in its expansiveness and its potential to fire the imagination. That said, when the first UELs arrived at Springhill, they didn't come here to seek adventures or to explore. Quite the contrary.

Under attack because of their allegiance to the British Crown, the UELs had abandoned their homes, lands, and possessions in the wake of the American Revolution. They'd fled north in the 1780s and early 1790s, carrying what little they could and hoping to build homes and carve out new lives for themselves and their families in a land the British had dubbed Cumberland.[4] With those goals in mind, the UELs set to work clearing acreages, erecting log cabins, and tilling the soil. In this area of the Cobequids, most of the settlers initially homesteaded on land a few miles west of the hilltop location that today is the Springhill townsite.

Relations between the newcomers and the Mi'kmaq, who'd been hunting and gathering in this area since time immemorial, were generally cordial. The Mi'kmaq traded meat and fish to the settlers in exchange for tea or tobacco. Other times, the Mi'kmaq were generous hosts who helped the newcomers survive by sharing their knowledge of which native plants were safe to eat and use medicinally, where best to fish and hunt, and how to tap into maple trees each spring to collect syrup. They also reminded the UELs of how easy and convenient it was to burn coal for warmth or cooking. Coal is the quintessential fossil fuel. It's also the world's most abundant energy source, being found on every continent— even Antarctica. People in China, ancient Rome, Europe, and the British Isles, as well as Indigenous Peoples here in Canada, have been burning it for thousands of years. As coal historian Barbara Freese has noted, "The industrial age emerged literally in a haze of coal smoke, and in that smoke, we can read much of the history of the modern world."[5]

On their seasonal fishing expeditions to the fossil-rich beaches at Joggins on the nearby Bay of Fundy, the Mi'kmaq found limitless quantities of coal. *Glmuejuapsgw* (gloo-moo-japs-gweh), they called it. Similarly, coal from the same seams was, and still is, found in abundance to the east, in the Cobequid Hills. In some places, the coal seams are so exposed and visible that you can find clumps of coal right under your feet. There's a good reason why.

Coal is sedimentary matter, the earliest of which began forming in the Devonian era, some 360 to 420 million years ago. That's when the first plants appeared. At that time, the land mass that today is Nova Scotia sat close to the equator, and the entire area that's now the Cobequid Hills was a lush swampland basin. The decomposition of dead trees and other vegetation—which geologists call "coalification," for obvious reasons—created thick layers of black, carbon-rich peat. As successive layers of sediments built up over countless millennia, the cumulative weight compressed the peat into lignite—"brown coal"—and eventually into the "black gold" that's real coal. None of this happened overnight. Far from it.

After many thousand years of accumulation, the decomposing plant matter produced three feet of peat. The increased temperature from this compression served as the catalyst for the chemical reaction that, tens of millions of years after burial of the peat under layer upon layer of sediment, resulted in the formation of coal.

THE FIRST FRENCH and English explorers who visited Nova Scotia in the sixteenth century found coal almost everywhere they looked. They found it in Pictou County, on Cape Breton Island, and in many other areas of Nova Scotia. The French began mining coal in Cape Breton as early as 1672, and in 1720 they started Canada's first commercial coal mine at Cow Bay. That colliery supplied fuel used in the construction of the Fortress of Louisbourg, which began the year before.

Coal was so abundant in Nova Scotia that in 1765, one ebullient British observer was moved to report that the mines in the colony consisted of "entire mountains of coal and are sufficient to supply all the British plantations in North America for ten centuries."[6] Small wonder that colonial officials in far-off London took notice when the British came to control all of Nova Scotia after booting out the French and moving the Mi'kmaq off their ancestral lands. Then, as colonial overlords were wont to do, the British set about exploiting Nova Scotia's natural resources. And so it was that King George IV granted his brother Frederick, Duke of York, exclusive rights to all the coal found in Nova Scotia (which at the time didn't include Cape Breton). Usually, the rich get richer. However, in this case the beneficiary of the king's largess was a sixty-three-year-old career soldier and wastrel with a fondness for playing cards and betting on horse races. And Frederick wasn't good at either. Perpetually in debt, he subleased the coal rights he'd been given to a syndicate of British investors, who formed an ambitious joint venture they called the General Mining Association (GMA).

The GMA brought the Industrial Revolution to Nova Scotia. The use of steam engines and railways helped increase the output from its mines, and experienced miners from the United Kingdom were hired to run those operations. As an article on the Nova Scotia Mining Association website explains, "While the GMA did many good things for Nova Scotia, its monopoly, and the heavy-handed ways it enforced it, were resented by many, including by other entrepreneurs prevented from pursuing mining opportunities."[7]

When in 1856 the GMA surrendered most of its mining rights, colonial administrators in London were quick to invite would-be entrepreneurs to apply for leases and subleases in the Nova Scotia coalfields. They did so in droves. "The immense and increasing consumption of [coal] for domestic use and for making steam in cotton and woolen factories, locomotives and steam vessels, and for the manufacture of gas and the working of iron, guarantees a demand for [the coal] in our very midst, which can only be limited by the supply and which cannot fail under judicious management to insure [*sic*] profitable returns," one enthusiastic investor gushed in 1864.[8]

In the thirty-five years between 1858 and 1893, more than thirty new coal mines were opened in Nova Scotia. Among them were the pits at Springhill. The presence here of coal had long been known; there are at least six main coal seams in the ground beneath the town. "The seams, separated by strata of sandstone and shale from thirty to three hundred feet thick, were once horizontal, but because they were raised by internal earth movement, slopes must be used to descend into the mines," local historian Bertha Campbell reported. "The seams dip to the northwest at an angle of thirty-five degrees."[9]

In the early 1830s, after surveying in the area, the legendary mutton-chop-adorned Nova Scotia geologist Abraham Gesner set about lobbying the colonial government to develop commercial mining operations at Springhill, River Hebert, Joggins, and other Cumberland County locales.[10] Geologists would subsequently find more than sixty coal seams in the Springhill area alone. Some were as thin as a couple of inches. Others were more than 14 feet thick. Not only was the coal abundant, it was readily accessible. "There was a time when men got coal out of their backyards," Bertha Campbell noted. "In recent years, there have been instances when a homeowner would step out of his door only to find a big gaping hole where his driveway had been. Another part of an old mine had caved in."[11]

As early as 1834, a Springhill settler named Lodewick Hunter set to work digging up bituminous coal that he sold to blacksmiths. Of course, it wasn't long before other entrepreneurs saw there was money to be made, and so they started putting shovels into the ground. Among the small-scale commercial coal mines that opened was one started by an enterprising man whose wife chanced to stumble over a clump of coal while she was walking in the woods.

Mining operations began in earnest in 1870, the year a trio of investors anted up the money to start the Springhill Mining Company. The No. 2 mine, which opened in 1873, was the oldest of five mines started near the townsite. It was also the most productive. No less significant was that after the mine shaft went deeper than 2,000 feet, No. 2 quickly acquired notoriety as the gassiest and most dangerous of Springhill's mines.

It was in 1873 that the Springhill and Parrsboro Coal and Railway Company opened a line linking the town to the port of Parrsboro, 30 miles to the southwest, and to Springhill Junction, four miles to the northeast. The former gave the owners of the Springhill coal mines access to lucrative, burgeoning markets up and down the east coast of the United States, while the latter provided a link to the newly established Intercolonial Railway. Beginning in 1872, it connected Nova Scotia to New Brunswick, Quebec, and Ontario.

The many thriving businesses along Springhill's Main Street were dependent on the DOSCO mine, which in this 1940s photo can be seen looming in the distance. (Nova Scotia Archives)

Springhill quickly became a boom town, one of Canada's most vibrant and fastest growing communities. In 1874, the population was two hundred. Within a decade, it was five thousand, and people just kept coming. Before long, Springhill had its own hospital, several primary schools, a high school, a horse racing track, a baseball field, a theatre that played host to well-known touring entertainers, and even a cricket pitch. Shops, restaurants, and businesses of all sorts lined the main street, which ran east to west downhill toward the colliery. Everyone and everything in town centred on the DOSCO mine.

With the North American economy humming, the market for Cumberland coal was insatiable. Springhill emerged as Canada's largest single coal producer in the 1880s. The mines were producing the cheap, abundant energy that the booming North American economy needed. Ol' King Coal was sitting high on his throne, and life in Springhill was good.

CHAPTER 2

THE BLACK GOLD
OF THE INDUSTRIAL ERA

THERE ARE FOUR VARIETIES OF COAL—ANTHRACITE, bituminous, sub-bituminous, and lignite. Anthracite, the highest grade, is hard and brittle; it contains a high percentage of fixed carbon and a low percentage of burnable matter. It's used mainly for heating.

Bituminous, which is the kind of coal that's found in the ground at Springhill and elsewhere in Cumberland County, is almost as hard as anthracite and has superior heating value. That means it can be used for both steel making and electrical generation. Sub-bituminous and lignite both have relatively low heating values and are mostly used in electrical generation.

In addition to the coal mines at Springhill, there were productive pits nearby at Joggins, River Hebert, and several smaller rural communities in Cumberland County. To varying degrees, all were going concerns. However, because Springhill's coal seams were bigger and of better quality, the mining operations there were the area's busiest, most productive, and most profitable. That made them a magnet for economic activity and investment. Sensing there was money to be made, in 1884 a consortium of Montreal industrialists and financiers swooped in to buy the Springhill mining operations. The ambitious new owners expanded the mines and amalgamated the two rail lines that carried coal to the port of Parrsboro and to markets farther afield in Canada and the United States. Going forward, the mines and railways were run as a single entity called the Cumberland Railway and Coal Company (CRCC).

The CRCC operated the Springhill mines until 1910. That year, Toronto-based Dominion Steel Corporation took over. The financial and corporate manoeuverings by British, American, and Canadian investors in the Nova Scotian coalfields around this time are too arcane and involved to delve into here; for our purposes, it's enough to say that in 1928, with the world on the cusp of the Great Depression, yet another corporate shuffle happened. The sadder-but-wiser British investors who'd sunk money into the venture created a holding company they called the Dominion Steel and Coal Corporation. It was DOSCO that would own and operate the Springhill mines for the next three decades while riding the ups and downs of the marketplace. Mostly downs.

The 1930s were lean times for Canada's coal mining industry, which was already in a long, slow downward spiral that began in the years after WWI; however, WWII brought a temporary spike in the demand for coal, the energy source that fuelled the wartime industries and military machines of Canada and its allies. During the six years of conflict, the Springhill mines employed more than thirteen hundred men. With jobs being so plentiful, the local economy prospered. "You couldn't find a parking space on Main Street on the weekends," one old-timer recalled. "Businesses were doing well, and people had money to spend."[1]

All things considered, Springhill was generally regarded as being a good place to live and to raise a family; with the town's crime rate being the lowest in Canada, the police force consisted of just three officers. There were sports leagues for the kids, nine busy, well-attended churches, and at least a half-dozen thriving fraternal organizations.

The good times continued for a few years in the postwar era, but Springhill's economic prospects gradually darkened. The boom years ended unexpectedly in the early 1950s. With oil emerging as the primary fuel for electrical generation and industrial uses, Canadian National Railway, Canadian Pacific Railway, and many of the regional carriers across the country, all of which had been important customers for Springhill's coal, stopped buying. Coal-powered steam engines were soon destined to go the way of the buggy whip and the corset. The shiny new diesel locomotives that replaced steam engines may well have been less photogenic, but they were cheaper to run and supplied more raw horsepower. That combination was unbeatable, and it was disastrous for the economic fortunes of DOSCO and for the one-industry town of Springhill.

Despite the fact the market for Nova Scotia coal was shrinking and prices were falling, in October 1957 DOSCO became the target of an unexpected hostile takeover bid. The suitor was an unlikely one—A.V. Roe Canada (known colloquially on Bay Street in Toronto as "Avro Canada"), the Canadian wholly owned subsidiary of A.V. Roe UK, which had been one of the world's first aircraft manufacturers when it was founded in 1910. In the early 1950s, Avro Canada was busy developing and building the CF-100 military fighter airplane and a supersonic jet interceptor that would become known as the CF-105 Arrow. The man piloting these initiatives was Crawford Gordon Jr. This forty-three-year-old Winnipeg-born industrialist, a flamboyant character if ever there was one in Canada's usually buttoned-down corporate board-rooms, was used to getting his own way in all things. Gordon also had an outsized fondness for alcohol and womanizing. "He had the kind of charm that just about dropped the pants off every lady he met," one journalist who knew him once observed.[2]

During WWII, Gordon had been one of "Minister of Everything" C.D. Howe's now legendary "dollar-a-year-men"—a team of top Canadian business leaders who volunteered to organize and direct Canada's war effort. Gordon had become president of A.V. Roe Canada in 1951, at a time when he was increasingly preoccupied with his efforts to get the Avro Arrow off the ground. This space-age aircraft was Gordon's pet project.

To help facilitate it, he restructured A.V. Roe Canada into two separate divisions, both of which were based at the Malton Airport in Toronto, now Lester B. Pearson International Airport. One division was known as Avro Aircraft Limited, the other as Orenda Engines. It was in a bid to diversify and expand Avro's business and grow its profits that Gordon initiated the purchases of several companies, one of which was DOSCO. Avro paid $60 million (about $600 million in today's money) for a controlling interest in DOSCO.[3] In retrospect, it was a move that was as puzzling as it was ill-advised.

If you go looking, you'll find few mentions of DOSCO and not one of its No. 2 Springhill mine in the various Crawford Gordon biographies or in the A.V. Roe corporate histories. Coal and steel production were low on the list of Gordon's workaday concerns. If they were at all important, it was only to the extent that they might generate revenue and help bankroll development of the Avro Arrow aircraft. Gordon seemed to acknowledge that his purchase of DOSCO was a tad risky when at the press conference where the sale was announced, he mused, "No doubt the coal industry is a worrisome picture, but we haven't studied this picture." Sir Roy Dobson, chair of A.V. Roe in the UK and the man who was Gordon's boss and mentor, was quick to add, "If we couldn't improve a company, we wouldn't go into it. If we can't improve DOSCO we'll be very disappointed people."[4]

It was against this backdrop that DOSCO management stepped up efforts to vertically align the ends of coal faces in the No. 2 mine. The hope was that doing so would result in increased productivity and profits while also reducing the incidence of bumping. However, there were a couple of major problems with this strategy. For one, increasing production of coal—or anything else, for that matter—at a time when prices are falling isn't a sustainable business model. For another, although the quality of Springhill coal improved as the mine went ever deeper, the ground around it became more unstable, and the work of the miners became increasingly perilous.

ACCORDING TO DOSCO records, the first seismic shock to rattle the No. 2 colliery happened in July 1917, when coal was being extracted at the relatively shallow depth of 2,000 feet. Over the next four decades, company officials and the inspectors from the provincial Department of Mines would record more than five hundred bumps of varying intensities. By the early 1950s, when mining was being done at ever greater depths, the number of bumps increased. Not surprisingly, so too did the sense of apprehension in Springhill.

In the wake of the fiery 1956 explosion that shut down the No. 4 mine at Springhill, the No. 2 pit was the last of the town's original quintet of mines that remained open. The quality of the coal being produced remained prime; however, prospects for the mine looked bleak going forward. The demand for coal was softening, and the mine itself was widely regarded as being "an accident waiting to happen." The overriding fear among the miners and their families was that when the next big bump occurred, it could close the mine and kill the town. The attitude of miner Gorley Kempt reflected how serious those fears were. Following a fire that razed the family home, the Kempts made do with whatever furnishings they could find. Gorley's son, Billy, remembers that his parents had no money for new items and his father refused to go into debt to buy any. He feared that if the mine closed, he'd find himself out of work and wouldn't be able to make the monthly payments.[5]

The coal company houses, drafty wood-frame structures in which many miners and their families lived, were scattered throughout the town of Springhill. (Nova Scotia Archives)

Gorley Kempt's fears—like those of many other Springhill miners—were well founded. Despite government subsidies, the No. 2 mine had become a money loser. Even more worrisome and threatening was the growing likelihood of a major bump. In the almost ten months of 1958 prior to October 23, there were eighteen bumps that injured forty-nine miners. In retrospect, one disturbance that happened on March 18 stands out, for it was eerily reminiscent of the Big One that would happen seven months later. One person died when "a violent bump occurred on the 13,400 level . . . heaving the pavement up one to three feet for a distance of one hundred and twenty feet."[6]

In large measure, the instability of the ground beneath Springhill stemmed from the Cumberland Basin's geological peculiarities; at least, that's how most of the geologists and mining engineers who studied the situation had it figured. But of course, there were other experts with different opinions. They were convinced the seismic disturbances at Springhill were triggered by the way mining was being done. That had certainly been the opinion of one George S. Rice, the chief mining engineer of the United States Bureau of Mines.

In 1923, provincial officials had invited Rice to inspect the No. 2 mine at Springhill and offer his recommendations on ways to reduce the alarming frequency and severity of the bumping. Rice was regarded as one of the world's leading authorities on coal mine bumps. After visiting the Springhill colliery and studying the data, he'd issued a report. His principal recommendation was that the "room-and-pillar" system of mining, the preferred method in the No. 2 mine for five decades, be changed to "longwall retreat."[7]

Room-and-pillar mining, which was in use in Europe as early as the thirteenth century, came to North America in the early 1800s. The basics of longwall retreat mining—which is also known as "the Shropshire method"—were developed in England in the late seventeenth century. While the technology involved in longwall mining has changed considerably, the basics remain the same. Miners remove as much coal as possible from a broad coal face and then allow the roof and any overlying rock to collapse into the void that's left behind, while maintaining (hopefully) a safe working space along the face for the miners.

The coal seam at Springhill was (and still is) a large sedimentary deposit that's several miles wide; the coal face had an average thickness of eight feet and as much as nine feet in places. Near the surface, the whole sheet slants at an angle of about 30 degrees. Deeper down—at a vertical depth of about a mile—the coal sheet gradually flattens out to about 20 degrees. Above and below are layers of sandstone, shale, and other smaller coal seams.

When coal mining began at Springhill, surface coal was the first to be removed, for obvious reasons. As mining continued and miners dug ever deeper into the earth, they turned to the room-and-pillar method of mining, removing coal by digging directly into the coal seam and carving out underground rooms, or stalls. The miners left uncut pillars of coal in place to support the mine's roof. "The thing is, when you're removing that coal seam, you have all that weight and pressure of the rocks above you all the way to the surface bearing down on the workspace, but you're also putting pressure right on the face of the coal you're working on," John Calder, senior geologist with the Nova Scotia Department of Natural Resources, explained in a 2018 media interview. "It acts like a lever, almost like a nutcracker, with the coal face being the nut, and the handles are the roof and the floor of the open space."[8]

After reaching the western-most boundary of the coal on each level in the No. 2 Springhill colliery, the miners began to work their way back toward the access slope, removing the coal in the roof pillars as they went. As this coal was extracted—as is the case in any mine—gravity dictated that the rock layers above would come down, sooner or later.

The art and the science of coal mining involve safely extracting as much coal from a coal face as possible. With that in mind, mining methods are designed in theory and then modified in practice according to the conditions dictated by Mother Nature—or Father Geology, as many geologists like to say (most of them being male). For instance, as the depth of the mining activity increases, the growing weight of the rocks and earth bearing down on the mine roof means the miners need to leave ever-larger pillars in place. This necessity makes it more difficult—and dangerous—to recover the coal in those pillars. If the roof layers don't break down easily, the pillars are prone to violent, uncontrolled cave-ins. Some coal is inevitably lost in the room-and-pillar method of mining, but depending on conditions, between sixty and ninety percent of it is recovered by this method.

Eventually, room-and-pillar mining becomes too dangerous and uneconomical. George Rice determined that Springhill had reached this point, and so he recommended a switch to longwall retreat mining. In theory at least, the actual method of mining coal in longwall is the same as for the room-and-pillar method: miners undercut coal along the width of a coal face. They then collect the coal as it falls. The difference between room-and-pillar and longwall retreat mining is in the pattern of extraction.

The word *retreat* is applied in longwall mining when a series of horizontal tunnels are driven off the transit slope out to the farthest limit of the area to be mined. At Springhill, the levels in the No. 2 mine extended for almost a mile.

The miners dug four levels of horizontal access with 400 feet between them. By October of 1958, the operating levels in the No. 2 mine at Springhill were at slope depths of 12,600, 13,000, 13,400, and 13,800 feet. The first longwall was started from the end of the 13,000-foot level by miners who extracted upward toward the 12,600-foot level. The connections between the levels were known as "heads." In effect, the head was a 400-foot portion of the coal face that was accessible from above on the 12,600-foot level and from below on the 13,000-foot level. Miners knew this work area as "the 13,000-foot wall."

As mining progressed, rows of wooden packs—support pillars that had been erected to control any fall of the roof behind the coal face—were removed, and the roof behind the work area was allowed to come down in a controlled way. Whenever all the coal had been extracted from the head that connected any two levels, work began on another one that was 400 feet deeper. That's how retreat longwall mining works in theory. How did it work in practice?

If you weren't claustrophobic, didn't suffer from nyctophobia (fear of the dark), and were brave enough to don a miner's helmet and headlamp to go on a sortie down into the perpetual blackness of the No. 2 mine, you'd have had quite an adventure.

Think of the mine as an italicized *E* that had been turned on its horizontal axis and given one extra prong. Above the top of the letter's vertical spine, off to the left, an older main access slope descended on a 30-degree angle from the pithead to a depth of 7,800 feet. At this point, there was a dogleg right. On your descent into the mine, you'd have exited the trolley car that carried you this far. You'd then have walked to a transfer tunnel that led to the top of the spine of that backward letter *E*. This was the top of the access the miners referred to as the "back slope."

After riding the trolley down the back slope, you'd have arrived at the entrance to one of the mine's four working levels—12,600, 13,000, 13,400, and 13,800 feet. (A level at 11,400 feet had been exhausted, and another at 14,200 feet was being developed, but it was never worked.) Let's say you descended to the 13,000-foot level. After exiting your trolley car, you'd have gone for another walk, this one to your right, in a westerly direction along the level until you arrived at the gob—the mass of waste material that's generated by the mining operations and the area where the roof had been allowed to collapse. *Gob* isn't a pretty name, but it fit. This area, the dead end on each level, served as an ersatz toilet and a garbage dump. Even though the mine was well ventilated, you'd have been wise to hold your nose and watch your step if you ventured here.

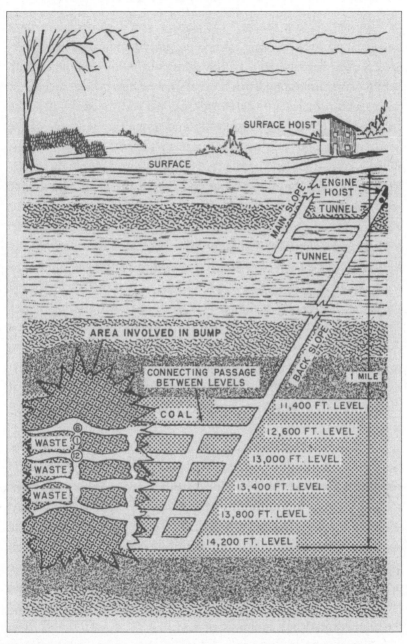

This simplified drawing of the layout of the No. 2 mine appeared in a 1960 study of the Springhill mine disaster. (US National Academy of Sciences–National Research Council)

Just before you reached the gob, if you'd turned your gaze upward and looked back over your shoulder, through the ever-present thin fog of coal dust that hung in the air, your headlamp would have illuminated the head that was mentioned earlier. At the 13,000-foot level, it extended 400 feet, all the way up to the 12,600-foot level. On your left, up and down the west side of the head, you'd have seen two or three rows of wooden supports—called packs—that held up the roof of the mine above where miners were working. Opposite, on the east side of the head, you'd have seen the coal face. The coal the miners chipped from it tumbled, or was shovelled, into long metallic pans that were joined together to form a chute. Gravity, with help from an engine that periodically rattled the pans, moved the coal downward to the bottom of the head. After cascading there, it fell onto a clattering chain-driven conveyor that moved it along the level to a transfer point, where labourers filled the coal-rakes that carried the coal out to the back slope and up to the pithead. Once it reached the surface, it was washed, weighed, processed, and readied for shipment. Some of it went by rail to the Bay of Fundy port of Parrsboro, while much of the rest travelled west via Canadian National Railway to customers in Quebec and Ontario.

As mining progressed and the coal face moved east and ever closer to the back slope, successive rows of packs were removed up and down the west side of the head, and the roof behind the work area was allowed to collapse. When this happened, the horizontal length of the level became shorter until finally the stock of minable coal was depleted and mining on that level ended. The miners then began working at a new level 400 feet deeper. Production in a mine never stopped, other than when there was a major bump or during the two-week period each summer when the mine shut down and the miners took their holidays.

This was the mining methodology in place at Springhill's No. 2 mine in the early 1950s. Much of the bumping that continued to occur when DOSCO introduced longwall mining, most notably from 1928 to 1933, was brought under control when the company closed off those parts of the coal seams that were especially unstable. That simple solution proved to be generally effective, and overall, the switch to longwall mining was beneficial for it reduced the incidence of bumping for almost three decades. However, all that changed when mining operations descended below 10,000 feet. Problems began to recur, and underground working conditions became more precarious than ever.

With the number of bumps rising again in 1953, provincial Department of Mines geologists began studying the situation. They set to work with DOSCO management to devise a long-term solution to the bumping problems. No one—not provincial officials, DOSCO management, the miners' union, or the people of Springhill—wanted to see the No. 2 mine's operations scaled back or, even worse, terminated. People tried not to think about that possibility, and they tried not to become prisoners to the fear that a catastrophic bump would likely kill a lot of miners and wreck the mine. Life had to go on, and it did.

The miners and their loved ones prayed during each shift that this wouldn't be the day the Big One happened. However, on October 23, 1958, Springhillers' luck ran out. And when it did, nothing was ever the same again.

CHAPTER 3

NO WORK, NO PAY

Thursday, October 23, 1958
2:00 P.M.

Harold Brine was one of the hundreds of Springhill miners whose life was about to change forever on this fateful day. But right now, it was a beautiful sunlit afternoon in late October.

Given the glories of the day and his own life circumstances, if you'd given Brine the choice of staying home today or going to his job at DOSCO's No. 2 mine, it would have been an easy decision for him to make. He'd have taken the day off. Unfortunately for Brine, he didn't have that choice. He couldn't stay home. He was scheduled to work the afternoon shift, 3:00 to 11:00 p.m., and there was no end of reasons why he had to show up. Most importantly, the twenty-six-year-old Springhill native needed money for the house he was building for himself, his wife Joan, and their chubby-cheeked two-year-old daughter. Little Bonnie was the apple of her dad's eye.

The Brine family's new home was on Mountain Road, a few miles southwest of Springhill. Brine still had a lot of work left to do on the place; it was a DIY construction site. The kitchen of the trim three-bedroom wood-frame bungalow was the only room he'd finished so far.

On this Thursday, the drywalling, plumbing, and painting had to wait. Just as he'd done every day of the seven years he'd worked for DOSCO, Brine tucked his company-issued lunch pail under his arm and hustled off to work. He was using every dollar he could muster to buy the lumber and sundry building materials to finish his house. At the same time, there were other bills to pay, and he had to put food on the table and buy gas for his car.

Regardless of how nice the weather was or how much he'd have welcomed a day off, Brine couldn't afford to stay home. Not today. Not any day. Doing so would have cost him as much as twenty dollars; that was a typical day's wages for him if he was able to exceed the production target and earn a bit of bonus money. But if Brine was ailing and stayed home, he didn't get paid. If he was injured on the job and couldn't work, it was the same deal. He didn't get paid.

Brine was a strong union man. He attended the regular Saturday "chapel meetings" of Local 4514 of the United Mine Workers of America (UMWA) at the Miners Hall whenever he could. He eagerly embraced whatever benefits his union reps could wring out of DOSCO, even though he and all the other miners knew the unalterable terms of employment at the Springhill mine. They were as simple as they were unyielding: no work, no pay.

Despite this reality, if he could have foreseen the hellish nightmare that fate had in store for him and his workmates today, there's no question Harold Brine would have stayed home. No amount of money could ever have prompted him—or anyone else who didn't have a death wish—to go to work in the No. 2 Springhill mine.

At the best of times, mining isn't a job for the meek, the frail, or the nervous. It never has been and never will be. It's understandable that many miners are on edge whenever they punch in. As a breed, they're superstitious. Every Springhill miner knew the story of the white witch from Miramichi who predicted mining disasters. Mother Coo was legendary. She'd correctly predicted explosions at two Nova Scotia mines. Then in early 1891, she'd warned that a blast would rock the mines in Springhill. Wary government officials, company managers, and union authorities carried out a thorough inspection of the mine, and then on February 19, they pronounced everything was good to go. It wasn't. Two days later, on February 21, the massive explosion that ripped through the No. 1 and 2 collieries killed 125 men.[1]

Small wonder that some miners were tormented by recurring premonitions of disasters. The angst gnawed away at them like a cancer. Harold Brine's co-worker Henry Dykens was among them. Fifty years old and already greying after too many years toiling as a coal miner, he had good reason to be fearful. Dykens had survived the 1956 explosion that killed thirty-nine men and brought about the closure of the No. 4 Springhill mine. The memory of that disaster had haunted him and many other survivors ever after. On this splendid October afternoon two years later, for reasons he couldn't pinpoint, Dykens felt especially edgy and unsettled. Gripped by a premonition of disaster, he dreaded going to work. Fortunately for him, he was fated to be one of the lucky ones who cheated death once again. Next day, he would tell a newspaper reporter, "I got up at 9 a.m. [Thursday] morning. I felt shaky all morning. I don't know why. I was still shaky when I went down [into the mine] at 3 p.m. Stanley Henwood was with me. He said, 'This place is due for a good bump.' I guess he was right."[2]

Dykens wasn't the only one who had the same fear. Far from it. Fifty-eight-year-old Ted Michniak had been a coal miner since 1918. Despite his forty years of underground experience, he never felt at ease on the job, especially after he started to work in the No. 2 mine. "I had a feeling of dread every time I went down there," he said. "That mine was too deep."[3]

ALBERT EINSTEIN WAS never a coal miner, but the renowned physicist might well have been thinking of coal miners when he mused, "Never think of the future. It comes soon enough."[4] Sage advice, that, and although Harold Brine wasn't even aware that he was doing so, he instinctively heeded it. He knew the dangers of his job all too well. Despite that, Brine never worried about what might happen when he was at work, in the perpetual darkness thousands of feet underground in "a pit filled with hell," as it has been aptly described. "I guess maybe I just didn't realize how dangerous working in the mine was," he said.[5]

Brine had never set foot inside a mine prior to the day in 1951 when he signed on to work his first shift with DOSCO. He was eighteen years old, full of youthful enthusiasm, fearless, and eager to start collecting a regular paycheque. Finding a full-time job was something that couldn't have come soon enough for him. Patience wasn't one of Brine's youthful virtues.

Born on June 25, 1932, the Springhill native was just eleven in April 1943 when his mother unexpectedly died after a brief hospital stay. Following Hattie Brine's passing at the still-young age of twenty-nine, her son and two daughters were raised by their father, Jerry Brine, with help from his parents.

It's impossible to say how much the loss of his mother affected young Harold's interest in—and tolerance for—sitting in a classroom. He was level-headed and plenty bright enough to do well academically, but he was impulsive and not inclined to sit still for very long. He also had a quick temper. Both those character traits and the fact that Harold was big for his age proved to be decisive factors in his life. So was his desire to grow up fast. He was in grade seven on the day in the autumn of 1945 that he decided he'd had enough of school. When a male teacher chided him for misbehaving in class, Brine sprang red-faced from his seat and dared the educator to "step out into the hall for a talk."

That challenge to do battle went unaccepted; however, it proved to be the coup de grâce for Brine's formal schooling. Understandably, the news that he'd quit didn't sit well at home. The Brines were hard workers, and so Jerry Brine demanded to know what his son intended to do now that he was no longer going to school.

"I guess I'm gonna have to find a job and go to work," he said.

"Damned right, you are!" Jerry Brine replied.

And so it was that at the ripe old age of thirteen, Harold Brine became a working man. He might well have followed in his father's footsteps and gone to work in the DOSCO coal mine then and there, had there been any job openings. There weren't. That winter, he had to make do working as a gofer—a "cookie," as locals called it—for a neighbour who ran a sawmill business; the pay was less than four dollars a day. After that, young Brine spent the next five years labouring at a variety of temporary part-time jobs. As you'd expect, none of them made him much money or had long-term prospects. The only real positive to come out of this period in Harold Brine's adolescence was that he got to meet and know a host of people around town. Everybody knew him, and he came to know everybody. "A lot of that had to do with me driving taxi for Gerald Henwood," Brine recalled. "I started doing that when I was eighteen. Gerald had an older Plymouth, and he was doing very well for himself, especially since he was bootlegging. He made enough to buy a new car."

Then, as now, bootlegging was illegal. That reality was driven home to Brine in no uncertain terms the night the town's eagle-eyed police chief, Leo MacDonald, stopped the cab Brine was driving while on a beer delivery. If he'd had a bank account, the resulting $200 fine would have put a huge hole in it—that is, if Gerald Henwood hadn't been kind enough to ante up. Regardless, the incident prompted Brine to look for another job.

By 1951, he was working full time at a Springhill garage and car dealership. Like cab driving, that job also provided no end of opportunities to meet people. When DOSCO's underground manager, Ronnie Beaton, brought in his car for servicing, Brine sensed an opportunity and seized it. Point blank, he asked Beaton for a job. Despite the mine manager's insistence that there were no openings at the mine just then, Brine pleaded. He told Beaton he was desperate for a better-paying job. He *really* needed more money. Brine told Beaton he and his girlfriend were planning to wed; truth be told, he admitted, the young lovebirds *had* to get married.

Beaton knew that Harold's bride-to-be, Joan Cormier, was a local girl whose father had worked at the mine. When Joan's parents split up, she and her sister had gone with their mother to Halifax, and there the Cormier girls attended a Roman Catholic convent school. When they returned to Springhill a few years later, Joan went to work as a waitress in Hyatt's Restaurant, a popular Main Street eatery. That was where she met Harold. Petite, dark-haired, and pretty, Joan was a couple of years younger than her handsome suitor. She was also shy and hadn't dated much. She and Harold were soon going steady. One thing led to another, and as has been known to happen when young people's hearts get the better of their heads, Joan found herself "with child."

Nowadays, almost thirty percent of Canadian babies are born to unwed mothers; it's no big deal to have a baby out of wedlock. However, social norms were markedly different in the 1950s, especially in the small towns of Canada's Maritime provinces. Values were conservative; people went to church, and social norms were strict. That meant "nice girls" didn't get pregnant. If they did, they were obliged to avoid public shaming either by getting married before the baby's arrival or by going away for a stay in a government- or church-run maternity home. When the "illegitimate baby" was born, the young unwed mother was pressured to surrender the child for adoption; most of these young women, who were vulnerable, traumatized, and ashamed, did as they were told.

Brine knew all this, and so did Ronnie Beaton. After confiding to the mine manager that Joan Cormier was expecting, Brine was quick to add that he was going to do "the right thing." He planned to marry her. Hearing this, Beaton agreed to do what he could to find a job for the young man. "I'll see what I can do for you," he said.

As it turned out, Joan wasn't pregnant. It had all been a false alarm. Despite this, Beaton was true to his word. Brine got a job at the mine. Then in early 1954, he and Joan married. By 1956, when their daughter Bonnie was born, Brine was twenty-four, had already been working underground for more than three years, and had settled into his job. As a new man in the mine, he'd earned $9.74 per day to start. The pay wasn't much, but it was steady, and it was enough that Brine was able to start building that house for himself and his family.

INITIALLY, WHEN HE signed on at the mine, Harold Brine had started "working on the timber," as his job was categorized on his employment contract. The fact this was "entry level" had no relation to the depth underground where the work was done. Brine was one of the legions of labourers who transported and helped install the wood used to build the elaborate wooden framework that supported the mine roof. Positioned every ten feet or so along the length of the horizontal level were vertical hardwood posts called "props." Each of them was about as thick as your leg and roughly ten feet high. These supported the "booms," the horizontal cross members that bore the roof's crushing weight. At the coalface, where mining was being done, the labourers stacked four-foot lengths of wood that served as temporary roof supports. These "packs" were removed whenever a section of the coalface was depleted and the roof in that area of the mine was brought down in a controlled collapse. It took a lot of time, effort, and wood to construct the elaborate wooden framework that kept the mine's roof in place while coal was being extracted.

Most of the timber work in the DOSCO mine was done at night, on what the miners called the back shift—from 11:00 p.m. to 7:00 a.m. This was also when mechanics fixed the machinery and moved the coal conveyor pans, trackmen maintained and repaired the mine's trolley tracks, and labourers cleaned up the levels where mining was done during the day shifts. The men who did these jobs were at the bottom of the workplace pecking order. By 1958, they were earning about ten dollars per shift. That was roughly two-thirds the salary of the men who did the actual hands-on extraction of coal. If there was an opening to advance, a man could "earn his papers" as a miner after two years' experience in the mine, as Harold Brine did. Doing so was worthwhile since miners were paid on a contract basis and they could maximize their paycheques by earning production bonuses; the greater the tonnage of coal they mined on a shift, the more they earned. "I can't recall exactly how much my pay increased when I moved from timber to coal, but I do know that it went up," said Brine.[6]

Minework was hard. It was dirty, and it was hazardous. Despite this, Brine enjoyed the camaraderie and the weekly pay packet. Each Friday, he lined up at the mine's payroll office to collect an envelope filled with his wages for that week. Like many miners, Brine never had a bank account. He turned his pay over to his wife, who looked after the family finances.

Joan hated that her husband worked in the mine, and she often told him so. She wished he'd find another job somewhere—especially after the 1956 explosion. However, steady jobs were scarce, as he was forever reminding her. Besides that, Brine said he liked working in the mine. It never bothered him to go down underground. He was fearless in that regard. The Brines were a typical Springhill family with mining in their blood. Jerry Brine had worked as a miner, as had his father before him. "Grandpa Brine was just eight years old in 1879, when he went to work in the mines," Harold Brine recalled.

Although the province of Nova Scotia mandated in 1883 that children between the ages of seven and twelve were required to attend school, it wasn't unusual for boys to skip school or quit entirely—as Harold Brine did—so they could go to work. In 1890, more than eleven hundred males under the age of eighteen were toiling in Nova Scotia coal pits. (It wasn't until 1923 that the province formally banned boys under sixteen from working in a mine.)

The youngest of the young coal miners took home twenty-five cents for a ten-hour workday; the older boys earned sixty-five cents. That money was hard-earned, no matter how old they were.

An article published in the December 4, 1890, edition of the *Halifax Morning Chronicle* described child labour in mines this way: "Long before your city boys are astir, the pit boy is awakened by the steam whistles, which blow three long blasts at half-past five o'clock every morning, thus warning him that it is time to get up. Breakfast partaken of, he dons his pit clothes, usually a pair of indifferent-fitting duck trousers, generously patched, an old coat, and with a lighted tin lamp on the front of his cap, his tea and dinner cans securely fastened on his back, he is ready for work. He must be at his post at seven o'clock. Off he goes, and in a few minutes with a number of others, he is engaged in animated conversation, and having a high old time generally, as he is lowered on a riding rake [a train] to the bottom of the slope."[7]

Boys performed a range of roles at mines, working both underground and on the surface. They opened and closed the ventilation doors that controlled air flow. They cleaned equipment. They sorted the freshly dug coal, and they cared for the pit ponies, the piteous beasts that hauled coal cars deep underground. The labour these boy miners did was physically demanding. It was hazardous, and it was hellishly dangerous. And all too often, it could be deadly.

Springhill mines were notoriously bump-prone and riddled with perilous volumes of the noxious gases that miners called fire-damp. Methane, the most prevalent gas, is an ever-present menace in any coal mine since it's a natural by-product of the same breakdown of plant matter that creates coal. Breathing in methane in low concentrations isn't lethal. However, if inhaled in high volumes or for a prolonged time, methane can kill or at least give rise to a variety of medical symptoms, none of which are pleasant. Even more problematic is that when the level of methane in the air in a closed space reaches 9.25 percent, the gas becomes flammable, and gas explosions can result.

In 1815, this danger prompted an English scientist named Sir Humphry Davy to invent a safety lamp for use in mines and other venues where flammable gas was in the air. A wire gauze chimney enclosed the flame in "the lamp that saved a thousand lives," as the miners often said, but the tiny holes in the gauze let light from the burning wick pass through it while the metal gauze absorbed the heat. This made the lamp safe to use in a mine since the flame couldn't heat enough flammable gas to cause an explosion. Almost as important was that the flame itself changed colour when methane was present, providing an early warning for the miners.

A more versatile alternative to the miners' lamp was a caged canary that alerted miners to the presence of methane and of the far more deadly carbon monoxide (CO) gas. It was in the late nineteenth century that British miners began taking caged canaries with them into the mines. The birds were chosen because the part of their nervous system that controls breathing closely resembles that of humans. Owing to the birds' rapid breathing, small size, and high metabolism, they were acutely sensitive to the presence of toxic gases, especially CO. Canaries are also renowned for their loud, melodious chirping. Miners carried caged birds—usually females, which aren't as melodious as the males of the species but were cheaper to buy—into the depths of the mine. The birds were not only cheerful company but also an early warning system; if they started to shake or stopped singing, miners knew there was methane or carbon monoxide in the air.

British coal miners stopped taking canaries into the mines in 1986, when low-cost electronic carbon monoxide warning devices became widely available. However, by that time the phrase "canary in the coal mine" had already entered our lexicon of idioms, and it's still used whenever we talk about "an early indicator of potential danger or failure."

MINING HAS ALWAYS been one of the hardest and most dangerous of occupations; even today, with safer workplace conditions, coal mining is still a hard way to make a living. In days gone by, accidents involving individuals or small groups of coal miners happened frequently. That was as true in Springhill as it was in other coal mining towns, and probably even more so.

In nine decades of commercial coal mining at Springhill—from the 1870s until 1970, when all mining finally ended—424 men and boys lost their lives in the town's coal mines. This news item, from the January 14, 1943, edition of the weekly *Springhill-Parrsboro Record*, was typical of similar stories that were published countless times over the years: "Another mine accident occurred on Monday in No. 2 mine; claiming the life of a young worker, Herman Weirwick, age thirty-one, while at his work. The accident occurred, when his arm, caught in a tugger rope which was running, was drawn over the drum inflicting terrible body injuries. Fellow workers stopped the engine and released the unfortunate man. While being rushed to All Saints' Cottage Hospital, he died on the way, in the ambulance. Following the accident officials held an investigation at the scene. Varley B. Fullerton, K.C., of Parrsboro, will hold the usual inquiry."

Although up-to-date, accurate figures are hard to come by, estimates suggest coal mining accidents worldwide still kill as many as twelve thousand miners each year.[8] A review of the list of Springhill mine fatalities includes the names of men who were killed by sudden, unexpected coal falls, struck by runaway trolleys, crushed between pieces of equipment, pulverized by explosions, or like the unfortunate Herman Weirwick, caught in a tugger rope. Some of these causes of death, while brutal, were quick. Others were slower and more painful; many weren't recorded on any official list. They happened years after a miner retired or left work with health problems. A miner's name wasn't chiselled into the edifice of the miners memorials on Main Street when he fell victim to black lung or any of the other chronic occupational hazards coal miners faced daily. Deadly though they were, these afflictions killed more slowly.

Canada's worst-ever coal mine disaster, which happened in 1914 at Hillcrest, Alberta, claimed the lives of 189 miners. However, that body count pales in comparison to North America's deadliest underground accident: a December 1907 explosion at a coal mine at Monongah, West Virginia, killed more than 360 men. And not to be macabre, but it's worth noting that the world's worst coal mine disaster occurred in 1942 at a mine in Liaoning, the northeast coastal province of Japanese-occupied China. Almost 1,550 miners died in the Benxihu mine, which was being run by the Japanese military. The safety of the prisoners of war and other slave labourers who were doing the digging was of no concern; the scale of the disaster reflected that grim reality.

Springhill's history up to 1958 included those two mining disasters that devastated the town and claimed 164 lives. The youngest of the victims was twelve-year-old Joseph Dupre, who was killed in the 1891 explosion. Incredibly, several boys as young as eight survived the blast and their exposure to lethal levels of firedamp. The other Springhill disaster—the one the aforementioned

Henry Dykens survived on November 1, 1956—resulted in thirty-nine deaths.

Harold Brine was painfully aware of the history of accidents and deaths in the Springhill mines, but like most of his workmates, he simply accepted the risks and shelved them in a back corner of his mind. As the miners liked to say, "The money was clean, even if at the end of a shift your face wasn't." It had been this way since Day One of commercial mining at Springhill.

IN 1958, THE AVERAGE Canadian blue-collar worker earned about $4,000 per year. The cost of living was relatively low at that time, especially in Nova Scotia and Canada's other Maritime provinces. It was possible for a typical family there to make do on a single income if the work was steady and the pay adequate. As a result, more than seventy percent of the men who worked at the DOSCO mine built and owned their own homes. Many of them also owned a car. Harold Brine certainly did. He'd bought, restored, and repainted a vehicle that had been damaged in an accident. His rebuilt maroon-coloured 1950 Meteor was a gleaming dream machine. The car, one of the earliest Ford models to have an automatic transmission, also had leather seats. And an AM radio. Whitewall tires. Tube lights on the fenders and lots of chrome. The whole nine yards. Small wonder it drew admiring stares from males and females alike.

Brine's spiffy wheels were a familiar sight on local streets, and so was he; Springhill wasn't a big place. Everybody knew just about everybody in this small town. People here married young. Most teenage girls took it for granted that when they came of age, they'd marry a miner, have babies, and (hopefully) live happily ever after. If Brine hadn't already worn a wedding band on his finger, the young single women in town would have regarded him as a prime catch. After all, he had a steady job, and he was good looking. There was no question about that.

Harold Brine behind the wheel of his souped-up 1950 Meteor. (Courtesy of Harold Brine)

Brine was average height and tipped the scales at about 175 pounds. He was built as solidly as a coiled spring; strong muscles were one of the few positive fringe benefits for the men who worked in a coal mine. Brine also was blessed with a winning personality and the kind of manly good looks that set feminine hearts aflutter. There was a touch of the Cary Grant twinkle in Brine's steely blue eyes. He surveyed the world from beneath dark eyebrows and a full, thick thatch of dark, wavy hair. Inexplicably, Brine didn't have the sideburns that were pretty much de rigueur for teen and twenty-something males at that time. Despite this, he combed back his locks in the same kind of ducktail coif that male heartthrobs James Dean and Elvis had made all the rage.

When the weather was warm and he wasn't working, Brine could often be seen motoring up and down Main Street. Seated behind the wheel of his shiny car, he well could have been a prototype for the *American Graffiti* hot-rodders who (in that 1973 film) were depicted as cruising the streets of cities from coast to coast across North America in the late '50s and early '60s. Brine's left elbow would jut out the open driver's door window, smoke trailing from the ever-present Player's cigarette that dangled from the corner of his lips. "I enjoyed myself. A lot of other guys cruised around town too. One of the boys had an old Model T Ford with a ragtop; others drove motorcycles. We all had a good time," said Brine.[9]

However, on October 23, 1958, he had no more time to cruise up and down Main Street than he did to work on his house. In order to be ready for the start of his shift at 3:00 p.m., Harold Brine made it his habit to be at the pithead a half hour before the mine whistle sounded. And so it was that at about 2:15 p.m., he hopped into his car with wife and daughter perched on the front bench seat beside him; at that time, there was no such thing as seat belts or baby seats.

As the Brines set off, driving northeast along Mountain Road, the afternoon breeze blowing in the open windows of the big Meteor was gloriously warm and summery. However, this late in the season the last leaves that still clung to the trees—stands of sugar maples, yellow birches, scrubby poplars, and staghorn sumac—along both sides of the two-lane blacktop were splashed with nature's autumn palette of colours. The reds, golds, and browns of the autumnal foliage stood in stark contrast to the cloudless blue sky above. It was a glorious day to be alive.

After a ten-minute drive, the Brines arrived in Springhill. At the foot of Main Street, Harold Brine turned his car to the left, passing through the iconic U-shaped archway that marked the entrance to the dirt road that led up to the lamp cabin building and the DOSCO mine. It was coming up to 2:30 p.m., and the dusty, gravelled employees' parking lot was already chock full of vehicles. Most were arriving; a few were leaving. The three o'clock shift change was just starting. The first of the men who'd worked the morning shift at the mine would soon be emerging from the depths, while Brine and the other men in the crews on the afternoon shift were checking in, getting ready to go below.

After giving Joan a quick peck on the cheek, Harold Brine hugged and kissed his daughter—"Bye bye, Daddy!" cried Bonnie—before he grabbed his lunch box, slid out of the driver's seat, and exited the car. Brine stood watching and waving a perfunctory goodbye as his wife wheeled the Meteor out of the parking lot.

"See you at eleven," Joan Brine called out through the open driver's window of the car. "Love you!"

CHAPTER 4

DESTINED TO BE MINER #872

—

THURSDAY, OCTOBER 23, 1958
2:20 P.M.

ON A WARM, SUNLIT DAY SUCH AS THIS ONE, THE THOUGHTS of many Springhillers often turned to the sport of baseball. Raymond "Tommie" Tabor was among them. Whatever he was doing, he'd sometimes stop and think what a splendid day it would be for a ball game. Today, as he did whenever he drove down Main Street on his way to work at the DOSCO mine, Tabor passed the Springhill ball field.

If you'd been in the car with him on this day, Tabor might well have pointed out the ballpark. Not that it was easily missed. The diamond was located at the bottom of the Main Street hill. It was there on the north side of the street, nestled on the infield of the town's horse racing track. Home plate was a long fly ball distance east of the decorative archway that marked the entrance to the DOSCO mine property.

It was no accident that the ball field occupied such a prominent location in Springhill. The town was baseball crazy. The local men's baseball team, the Fencebusters, and their exploits were—and are—

the stuff of local sporting legend. In the twenty-four years from 1921 to 1945, the team made it to the provincial finals sixteen times, winning eight provincial championships. In the days before every home had a radio or television set, whenever the Fencebusters played a big game on the road, people followed the action via telegrams that a reporter wired back to Springhill after each inning of play. Large crowds would gather to read these reports, which were posted in the window of the telegraph office at the Springhill railway station.

Many of the players who suited up for the Fencebusters were miners living out their diamond dreams. As many as three thousand people—almost half the town's population at that time—sometimes packed the bleachers on a Saturday afternoon to watch them play. As a boy, Tommie Tabor was often there among the throng, cheering on the home side. Then, in his teenage years, he played for the town's junior men's team. Tabor was a gifted athlete, a "take charge" guy who was blessed with a strong, accurate throwing arm. He'd been the catcher on the 1937 Springhill Red Sox team that thrilled all of Cumberland County by winning the junior men's championship of the Maritimes. He'd graduated to play with the Iron Dukes intermediate team and then with the Fencebusters.

As you may have guessed, the Springhill Red Sox name was a nod to the major league's Boston Red Sox, which had a large following locally. Fans tuned their radio dials to AM 850, Boston station WHDH, to hear broadcasts of the Sox games. If you'd walked around the residential streets of town on a lazy summer evening, chances are you'd have heard the radio broadcast of a game drifting out through the open windows. When the Red Sox were playing in the afternoons, miners working the 3:00 to 11:00 p.m. shift at the mine would gather outside the nursing station to listen to the radio broadcast of the game until it was time to go below. As often as not, you'd have found Tabor among the men who were there listening. He lived and loved baseball.

Life had thrown Tommie a called third strike that curtailed his own playing days. Even if he'd seen it coming, there probably was nothing he could have done to change the pitch's trajectory.

There are those who believe and insist it's fate that dictates the life paths we follow. Maybe they're right, or maybe they're not. Who can say for sure? Regardless, it looked as though kismet had decreed that Tommie Tabor would earn his living as a coal miner. He'd been born into a Springhill coal mining family, and it seemed he was fated to follow in his father's footsteps.

Tommie's dad, Percy ("Pa") Tabor, was twenty-four in 1895, the year he began working underground at the only job he'd ever know. Fittingly, when Percy married in 1902, his seventeen-year-old bride, Myrtle Johnson, also hailed from a coal mining family. By the time Percy retired in 1947, he'd laboured in the pits for fifty-two years and had left his flesh-and-blood legacy there.

Among Myrtle and Percy Tabor's ten kids were seven sons. Raymond Tabor, who entered the world on February 29, 1920—as a leap year baby—was the couple's seventh child. It was his older sister Margaret who dubbed him "Tommie." He liked the nickname, which was matey and suited him. He was a people person. He made friends easily and was never shy about getting involved in the community.

All seven Tabor brothers toiled in the mines at one time or another. For three of them, mining was fated—there's that word again—to be their life's work. In the autumn of 1958, Tommie was thirty-eight, and he'd already been a miner for almost half his life. That being so, he knew all too well the dangers of underground work. It would have been impossible for him not to. Accidents in the DOSCO mine had claimed two of his siblings. Tommie Tabor's older brother Fred died in a 1938 rock fall; he was thirty-one when a boulder suddenly dropped from the roof of the No. 2 mine and struck him on the head. A younger brother, Donald, fell victim to carbon monoxide poisoning in the 1956 disaster that shattered the No. 4 mine.

Most coal miners seldom, if ever, voiced their vocational fears.

Oh, sure, they sometimes joked about them, or they'd raise safety concerns in the weekly meetings held each Saturday at the union hall, but that was about it. There was no point in dwelling on any of this stuff. The wife of a miner who died in a mining accident put it succinctly and well when she said, "We never talked about the risk. It was just something that we lived with."[1]

That was Tommie Tabor's attitude. And so, while he never talked much about the deaths of his two brothers, there can be no doubt that he felt their loss deeply. Sometimes late at night, when he and his wife, Ruth, lay in bed talking about the latest goings-on at the mine—accidents, close calls, or deaths that had happened—Tabor would begin to shake. Ruth understood why. She was a McManaman, a well-known Springhill mining family; three of her brothers worked alongside her husband at the DOSCO mine. By 1958, after almost two decades of marriage, Ruth Tabor well knew how her Tommie felt about working in the No. 2 mine. She also knew that if there'd been a good option, he'd much rather have earned his living another way.

Tabor had been just seventeen when he'd joined his father and brothers on the payroll at the DOSCO mine in 1937. It was pretty much a given that he'd do so. He was already a veteran miner when he and Ruth McManaman got married in November 1941. The newlyweds moved in next door to his parents in the vacant half of the coal company house at 86 Herrett Road. Ruth Tabor and baby daughter Glenda remained there, close to her in-laws, when Tommie went away during WWII, serving for three years and fighting overseas as a member of the Royal Canadian Corps of Signals. On a per capita basis, Springhill had the highest enlistment rates of any town in Canada. Even though the staff at the local recruiting office posted a sign that said, "If you are a Springhill Miner you are needed in the mines, not in the army," more than 1,250 local men and women signed up. Almost seven hundred of them served overseas, and fifty-four of them were killed.[2]

When he wasn't wearing a baseball uniform, Raymond "Tommie" Tabor wore a different kind of uniform—
that of a soldier. He served three years in the Canadian Army during WWII. (Courtesy of Valarie Alderson)

FOLLOWING HIS RETURN home from Europe in early 1945, Tommie Tabor resumed his life in Springhill pretty much where he'd left off. Like the million other Canadian men and women who'd been in uniform during the war, Tabor wanted nothing more than to settle down, raise a family, and enjoy the peace and prosperity for which he'd fought. His workaday routine in Springhill fit him as comfortably as his old catcher's mitt. If Tabor's life was remarkable in any way, it's that it was so typical of the life of the many Springhill miners who'd volunteered to do their bit for King and country

Although he went back to work as a miner in the No. 2 mine, it was Tabor's family life that was his focal point. Tommie and Ruth were devoted parents. In addition to daughter Glenda, the couple had three more children. In 1946, Ruth gave birth to a son they named Gary. Three years later, daughter Valarie came along, and then in 1953, daughter Susan—"Suzie Bubblegum" her dad called her.

Even as he was working full time and engaged in a busy family life, Tabor somehow found—or made—time to involve himself in the community. He volunteered with the Knights of Columbus, played some baseball with the Fencebusters, served as president of the local women's softball league, and coached both a women's team and his son Gary's Little League ball team. Small wonder the Tabors' company house on Herrett Road was perpetually buzzing. There were always four kids and their playmates running around, of course. Then too, with the DOSCO mine being close by, Tommie's pals were always dropping by before or after work, and just a few doors down the street was "the Liars Bench," a neighbourhood gathering spot for miners young and old alike who were tellers of tall tales. Pa Tabor had gathered scrap lumber and helped build and set up the rough-hewn bench that became—and still is—a Springhill landmark.[3]

A steady stream of visitors came to the door on both sides of the Tabor duplex. On Tommie and Ruth Tabor's side, the kitchen was a prime draw. Ruth, like her husband, was quiet by nature, but she loved to cook. The pies she baked spoke on her behalf with the eloquence of a great orator. "Butterscotch, lemon, coconut . . . I try to have something different every day," she once told American author Melissa Fay Greene.[4]

In 1942, when Pa Tabor and his chums built a bench adjacent to the Tabor family's home on Herrett Road, pranking teenagers painted the seat and dubbed it "the Liars Bench." The name stuck and the Liars Bench became a Springhill landmark. (Nova Scotia Archives)

Visitors also called next door, at the home of the elder Tabors, Myrtle and Percy. "A lot of men used to come to talk with my grandfather about what was going on in the mine," Valarie (Tabor) Alderson remembered. "Grandfather had a little shack [in the yard] where he and the men would go to play cards, smoke, and have a few drinks. Mum Tabor wouldn't let them do any of that in the house."[5]

Each July, when the DOSCO mine shut down for the annual two weeks of holidays, Tommie Tabor always made it a point to take Ruth and the kids on a road trip. They'd pack up the family's 1954 Ford and head for the seashore. Some of their happiest memories were made on the drives they took around Nova Scotia and over to Prince Edward Island. One of the most cherished photos in the Tabor family album is of Tommie and his two younger girls, taken in the summer of 1958 in the passenger lounge of the Northumberland

Strait ferry. For Tabor, life was always all about family. That was why when Ruth's aging parents began having health issues and could no longer look after themselves, Tommie, Ruth, and the four kids moved in with the McManamans in their house on Pleasant Street. This was a real change for the Tabors. They became "hillers."

This photo of Tommie Tabor and his daughters Susan (*on her dad's lap*) and Valarie on the ferry to PEI is a cherished family keepsake. (Courtesy of Valarie Alderson)

Most of the residents along Herrett Road, on the flat land to the west of the town that was a hop, skip, and jump away from the mine, were miners and their families. Pleasant Street, a half-mile to the east on the hilltop, had a different focus. Not all the male family breadwinners here were miners. The neighbourhood was known as "French Hill" since so many families with French surnames lived there. Not that any of this mattered much to the Tabors. After their move, the only real differences in their lives were that the kids now went to a different school and not as many familiar faces dropped by unannounced for a cup of tea or to sample Ruth's pies. For her husband, the biggest difference was that while he could walk to work in five minutes when he and his family lived on Herrett Road, he now drove most days. That's how, on this balmy autumn afternoon, he came to be passing the Springhill baseball field. He'd snatched up his lunch box and identification tag, given Ruth a peck on the cheek, and ruffled Suzie Bubblegum's hair as he breezed out the door; the other three kids were still at school until four o'clock.

Tabor's drive to the mine was a short one—northwest to the end of Pleasant Street, hang a left, and then motor straight down Main Street hill and past the ball field. It would take him ten minutes, tops. There'd be lots of time for him to change into his work clothes, pick up his headlamp, and have a smoke before going down into the mine. All things considered, today seemed like a fine day, and tomorrow would be even better. It would be Friday, and that was payday.

CHAPTER 5

THE SINGING MINER

Thursday, October 23, 1958
2:30 p.m.

WHAT QUALIFIES A PERSON TO BE REGARDED AS A HERO? That's a great question. The dictionary advises us that a hero is someone "who is admired or idealized for courage, outstanding achievements, or noble qualities."

Heroism, like beauty, is in the eye of the beholder, of course. But if you accept the dictionary criteria, it's clear the man who was fated to become one of the iconic figures in the Springhill mine disaster of 1958 was a hero, albeit a most unlikely one. He never aspired to be a hero, nor did he ever claim to be one. The label was one the media applied to him, and it stuck because it fit.

Maurice Ruddick was an unassuming man of colour who in many ways was the quintessential Canadian. Soft-spoken, polite, and compassionate, he was unfailingly upbeat even when he had every reason not to be. And at age forty-six, Ruddick was a devoted family man who loved people and the things in his life that truly mattered to him. Small wonder that even though six decades have passed since the disaster that shattered the Springhill mine, many Canadians—especially Maritimers—still remember Maurice Ruddick with fondness and no small measure of reverence.

What people recall and most admire about him is that he was one of those rare people who display grace under pressure and without hesitation or second thought do whatever it is that needs to be done. In today's world, such people are rarer than monarch butterflies.

We all like to think that in a time of crisis, when it's all on the line and lives may hang in the balance, we'll find within us whatever it takes to stay calm, see the bigger picture, and think about the needs of others as much as our own. If Ruddick was renowned for anything prior to the Springhill mine disaster, it wasn't public service or civic engagement. No, it was his love of music. That's what defined him in the eyes of many people in Springhill. And there was a good reason why.

There's a familiar old Disney movie song that advises "whistle while you work." Ruddick gleefully embraced that credo. Music gave form and direction to his life, and it was music that would in their darkest hours offer Maurice and his co-workers hope when they needed it most.

To some people, he sounded like Jimmy Rogers, the well-known American bluesman of the 1950s. Others heard in his voice a hint of the nasal twang of Hank Snow, the Nova Scotia–born country and western singer who was one of the province's favourite sons. Such comparisons were all well and good, but the truth was that Ruddick was his own man. His vocal style was as unique as it was varied; he sang the blues, pop tunes of the day, what were known as "old Negro spirituals," and the ballads he crafted. Although he was a natural tenor, Ruddick could croon as a baritone or bass when he felt like it.

At DOSCO's No. 2 mine, this self-described "coal mining daddy" delighted in entertaining himself and his co-workers. It would be quicker to talk about where Maurice didn't sing than where he did. He sang while riding the trollies that carried the miners down to the coal faces and up again after their eight-hour shifts. And he sang while labouring shirtless, dripping with sweat, and coated with the coal dust that inevitably eroded miners' lungs and killed so many of them.

Ruddick also sang as he strolled along the streets of Springhill. He and three pals sang in a quartet. He sang in the choir at the local Baptist church where he, his wife, Norma, and their children never missed a Sunday service. And he sang in the half-century-old coal company house on lower Herrett Road that he and his family called home. Ruddick sang whenever he got the chance. He sang at the kitchen table, and he sang as he was tucking little ones into bed. On those rare occasions when all was quiet—and, mind you, that wasn't often in a crowded three-bedroom abode—Ruddick would strum his guitar and croon into the microphone of an old reel-to-reel tape recorder he'd picked up somewhere. He loved to sing, and music was one of the few joys in life that was free.

Money was always tight in the Ruddick home. It was so tight that the night Maurice lost his dentures down the hole in the out-house, he had no choice but to use a stick to retrieve them. He knew there was no money for new ones. He also knew if he did raid the family's piggy bank, it would mean less food on the table, or maybe one of the kids would have to go without a new pair of shoes or a winter coat. And so, after boiling his dentures thoroughly in hot water and soaking them in bleach, Ruddick reluctantly popped the choppers back into his mouth. Very reluctantly. "Next day, when he told Mom what had happened, they both laughed about it," his daughter Valerie (Ruddick) MacDonald recalled. "But Mom always said that for weeks afterward, whenever she kissed Dad, she thought she could smell poop. She knew she couldn't, of course, but it was the thought of where those teeth had been."[1]

As this incident attests, while Maurice Ruddick was a proud man, he had a lively, at times self-deprecating, sense of humour.

Despite having to endure indignities and forever struggling to make ends meet, he always looked on the bright side of life. Doing so enabled him to love his job even though he was acutely aware of the dangers of minework. It was that awareness that prompted Ruddick to take the day off whenever it was his birthday, even though it meant the loss of a day's pay. His rationale for this sacrifice was simple: he didn't want to "ruin his special day" by dying in a workplace accident. He once told an interviewer, "If I could get a driver, I'd celebrate." Then, after a short pause, he added with a mischievous laugh, "Moderately."[2]

Ruddick always worked hard, never complained. Nor did he ever shirk family responsibilities. On the contrary, each weekend he somehow came up with money to buy his kids hot dogs, ice cream, or some other treats. And whenever he could, he took Norma to play bingo or they went out dancing. Ruddick had always wanted to have a big family, and he celebrated when he did.

He'd caught the eye of his wife-to-be, Norma Reid, the moment her older sister, Stella, introduced them. "I was eighteen when I met Maurice. It was love at first sight," Norma would always remember.[3] Both of them were natives of Joggins, the historic Bay of Fundy mining community that's located 25 miles west of Springhill. The Ruddick and Reid families had been neighbours, and Ruddick, who was thirteen years Norma's elder, never forgot that it was on a visit to the Reid home that he'd first seen her.

Not long after Norma and Maurice wed on November 30, 1945, they had occasion to make wishes over a turkey wishbone. When Maurice came away with the bigger half, Norma was curious enough to ask her husband what he'd wished for. Maurice grinned sheepishly. "I wished we'd have twelve children," he confided. He wasn't kidding.[4]

By 1958, the couple had that dozen children, and Norma Ruddick was busier than a ticking clock. Every minute of every day.

Norma and Maurice Ruddick. (Courtesy of Valerie [Ruddick] MacDonald)

Even when the eldest children were old enough to begin helping with household chores, it was no small feat for Norma to keep the house clean, do the washing, change diapers, sew all the frayed clothing, soothe hurts, and cook meals for herself, husband Maurice, the children, and her elderly father-in-law, a retired miner who was an integral member of the family until he died in April 1958.

Apart from the wife and children he adored, music was the one other treasure Maurice Ruddick had in his life. And music became his gift to each of his children and to the people around him. It would prove to be a lasting gift not only for his children but also for others who never would have expected that to be the case.

Ruddick taught all his kids to sing and play musical instruments. Three of the Ruddicks' eldest daughters—Sylvia, Valerie, and Ellen— were blessed with full, beautifully rich voices. They performed alongside their father in people's homes, in church basements, at funerals, at lobster festivals, and at the Miners Hall. They sang anywhere they could. Not only did the Ruddick family ensemble sound good, the girls looked smart in their outfits—navy blue with white polka dots—which were the first new store-bought dresses the girls ever owned. "Dad billed himself as 'the Singing Miner,' and we were 'the Minerettes,'" Val MacDonald remembered.[5]

When the girls inevitably began to spread their own musical wings, they moved to Halifax in hopes of launching professional singing careers. This was something Maurice himself had always dreamed of doing; if only his life circumstances had been different and he'd had the opportunity, he very well might have had a professional singing career. But he never got that opportunity. Mind you, there were compensations. As Ralph Waldo Emerson once said, "For everything you've missed, you've gained something else, and for everything you gain, you lose something else." That was certainly true for Ruddick.

While he hated to see his girls leave home, he was proud of them and prayed that music would be their ticket to a better life. As for himself and the rest of the family, Ruddick accepted that life would go on. It had to, and it did. Ruddick had taught all his children to harmonize. Subsequent iterations of the Ruddick family chorus billed themselves as "Maurice Ruddick: the Singing Miner and the Harmony Babes."

It was inevitable that many people in Springhill referred to Ruddick as "the Singing Miner." However, for those who didn't know him, there was another aspect of Maurice Ruddick's identity that defined him. To them—and to no small degree, to Ruddick himself—that was his skin colour.

———

MAURICE ALBERT RUDDICK was born September 13, 1912. His father, G. Ernest Ruddick, was a Joggins coal miner. His mother, Esther (née Issard) had worked as a domestic. Young Maurice's skin tone was dusky olive; he "looks as if he might have a little Latin blood in [him]," one Toronto newspaper reporter noted.

Had he chosen to do so, young Maurice likely could have passed himself off as being white. However, that's not what he was, nor was it how he saw himself. "I'm a Negro," he once told that same reporter. "I'm not a thoroughbred. I'm somewhere over half. If you're half, you're usually neutral, but I'm somewhere over half and so is my wife."[6] Ruddick proudly self-identified as being "mulatto" or "coloured" (the latter being the term many Nova Scotians in the 1950s used when they talked about Black people).

Maurice Ruddick was a man of quiet dignity. He was soft-spoken and God-fearing. Unlike so many of his co-workers at the mine, he'd finished high school and could read and write well. That wasn't all—other than his race—that set him apart from most other coal miners. He never cussed. He smoked only an occasional cigar, drank sparingly, and was unfailingly polite—at times to the point that strangers assumed he was shy. Then too, there were those bigoted people who simply figured he was a Black man who "knew his place." Regardless, it was evident to one and all that Ruddick had an undeniable presence. "Maurice was the kind of person you'd say could make friends with anyone. If you didn't want to speak with him, you still had to speak to him," Norma Ruddick once explained.[7] At the same time, she knew her husband was no shrinking violet.

Ruddick was of average height, and like many coal miners he was muscular and barrel-chested; that latter trait probably had something to do with his large vocal range. When it came to his personal appearance, he took great pride in how he looked. *Debonair* was the adjective that sprang to mind for many people when they thought of Ruddick. Why not? The descriptor was apt.

He parted his wiry, pomaded hair down the middle in a style that was in vogue back in the 1920s; however, it still suited him, and it looked good. Ruddick's face was broad, his jawline strong and dimpled. His dark, full eyebrows framed sensitive eyes that reflected the same quiet intensity that came out in songs and poems he wrote. The pencil-thin moustache that adorned his upper lip exuded a devil-may-care aura—à la 1940s Hollywood film stars Clark Gable and Errol Flynn. Ruddick's taste in clothes ran in a similar direction.

He favoured bow ties and was fond of brightly coloured plaids—maybe that was a reflexive nod to a rumoured hint of Scottish blood somewhere deep within his veins—even when the tartan patterns of his pants and jackets clashed. In warm weather, he was often seen strolling along the streets of Springhill looking sportily dressed. A stylish felt fedora, which he liked to wear at a jaunty angle, topped off his *au courant* look. "When you go around town, hold your head up, and walk proud," he advised his children.[8] Mind you, at that time doing so wasn't always easy for a Black person in Nova Scotia or elsewhere.

Although, in the late 1950s, the civil rights movement was gaining momentum both in the United States and Canada, racism was still very much a fact of everyday life. In Nova Scotia, the experiences of Viola Desmond and of the residents of the Halifax neighbourhood known as Africville underscored that ugly reality. Nova Scotian civil rights activist Verna Thomas once wrote: "The colour bar meant that Black people were barred from everything except the barnyard. We were certainly barred from opportunities for employment. I had the surprise of my life the day [in 1956 that] I sought employment at Moirs' Candy Factory [in Dartmouth]. When I went to apply, they asked me if I was Black. . . . When I answered 'Yes,' they told me that they didn't hire Black people. Leaving there, I went to the Metropolitan Store and spoke to the manager. . . . He told me there weren't any jobs available, yet there was a 'Help Wanted' sign posted in the window."[9]

Thomas's experiences weren't unusual. In the middle decades of the twentieth century, some historically challenged whites in Nova Scotia continued to treat Blacks as if they were unwelcome newcomers. The reality was there'd been a Black presence in Nova Scotia for almost three hundred years. The first person of colour to arrive was Mathieu da Costa, an interpreter in the employ of the French explorer Pierre Dugua de Mons, who visited Port Royal around 1608. The first Black settlers who came to Nova Scotia to stay were among the United Empire Loyalists who arrived in the years after the American Revolution (1775–1783).

Roughly two-thirds of the fifty thousand UELs who made that move put down roots in Nova Scotia and Prince Edward Island (which at that time was known as St. John's Island). Among them were twenty-five hundred Blacks, about five hundred of whom reportedly were "free" and came north to take up the British promise of "Freedom and a Farm." Sadly, that British pledge was never fully honoured. As a result, more than sadder, but wiser, Black UELs—including Rev. David George, Canada's first Black pastor—left Nova Scotia for Sierra Leone in 1792.

It was around this same time that white UELs began arriving in the Springhill area. According to local historian Pat Crowe, Springhill's first Black resident is believed to have been a man named Moses George, who probably arrived in the early years of the nineteenth century. Odds are he was a former slave, although that too is just speculation; nothing certain is known about him. The next Black man to arrive at Springhill was twenty-nine-year-old John Izzard, who worked in the mine. In the late nineteenth century, the town's collieries were at peak output and employed more than fifteen hundred men. Springhill was a magnet for job seekers of all ethnicities.

It was the offer of a chance to work on the back shift—

11:00 p.m. to 7:00 a.m.—at the DOSCO mine that prompted Maurice Ruddick's move to Springhill. The 1939 outbreak of war in Europe had breathed fresh life into mining operations at Springhill after the lean years of the 1930s. The demand for coal to power Canada's industries, railways, and navy was insatiable during WWII, and so Springhill's population ballooned, and its economy flourished anew.

During this period and into the late 1950s, more than ninety-five percent of Springhill residents were white Anglo-Saxon Protestants. With WASPs outnumbering Catholics by ten to one, the Orange Order was a going concern in the town; it had been since 1873.

There was very little identifiable French presence, and other ethnic groups including Jewish, Chinese, East Indian, and Indigenous Peoples were as scarce as sunbeams in a mine shaft. A small Mi'kmaw community, consisting of just a few families, lived at Springhill Junction, four miles to the northwest. Otherwise, most of the Indigenous Peoples whose ancestors had inhabited this area for thousands of years were relegated to reserves allotted to the Pictou Landing First Nation, 90 miles to the east.

Springhill's Black population was small, about twenty families for a total of around two hundred people; several of the families had double-digit totals of children. While there was no Black neighbourhood per se, several Black families were neighbours of the Ruddicks on lower Herrett Road, a stone's throw from the DOSCO mine. Sadie (Ashe) and Fidell Allen and their four children lived in the house next door, while Bertha (Hatfield) and Bob Silvea with their fifteen kids, and Florence (Dorrington) and Tom Gabriel with their nine, all made their homes in nearby company houses. Other Black families lived at Miller's Corner, a few miles west of town. The fact there weren't all that many people of colour in Springhill doubtless explains why the Black community was so tightly knit.

With apologies to American political scientist Lee Atwater, who advised that "perception is reality," in small-town Nova Scotia— and just about everywhere else in the Canada of the 1950s—Black people and white people experienced very different realities. They also had very different hopes and expectations. Black people saw things differently because each day they had no choice but to live with the cold, harsh impacts of racism and white privilege. Norma Ruddick said it succinctly and well in 1997 when she told inter- viewer Melissa Fay Greene that white people in Springhill simply didn't—and they couldn't possibly—know about or understand the racial prejudices that people of colour faced. Such things were never discussed. "It's a lot better now than it was," she said. "But I don't think it'll run out because you'll find it somewhere along the line."[10]

Despite this racism and despite the town's demographic pro- file, Springhill wasn't a bad place for Black people to live in the 1950s. White and Black children attended school together—unlike the situation in some parts of Nova Scotia, where school segrega- tion was very much a reality.

Springhill children of different races also played together. They sang together too. Anne Murray, who was thirteen at the time and destined to enjoy an international singing career that would win her acclaim as Springhill's most famous citizen, knew all about that. She chummed with Ellen, Sylvia, and Valerie Ruddick. "They were close to my age and my brother Bruce's age, so we had them over to our house all the time," said Murray. "We sang, and sang, and sang, for hours and hours. The Ruddick girls were wonderful. I learned a lot from them. It was a revelation when they introduced me to gospel songs by Mahalia Jackson and other great gospel singers."[11]

UNLIKE THEIR CHILDREN, the adults in Springhill's white and Black populations tended not to mix socially. Not much, anyway. It was a different story at the mine. Virtually all the town's Black breadwinners worked at the DOSCO mine. There they toiled elbow to elbow with their white co-workers. By the autumn of 1958, Maurice Ruddick had been doing so for more than seventeen years. This meant he was one of the veterans at the mine. Fifty-eight-year-old Ted Michniak was the dean of the crew in which Ruddick normally worked, and the two got along well, sometimes even sharing their lunches; one of the youngest in the crew was twenty-six-year-old Harold Brine, the same miner who drove the souped-up 1950 Meteor that was a familiar sight on Springhill's Main Street. Brine had started working at the mine when he was nineteen, and by 1958 he had about seven years in. He got along with everyone, regardless of skin colour.

"The Black people I worked with all were good. I never had a problem with any of them, and I hope they didn't have any problems with me," said Brine. "When I first started in the mine in 1951, I got into an argument with a Black fella I was working with, and I said to him, 'I'll take a round out of you!' He looked at me and said, 'You couldn't throw sand in my face.' We both laughed, and that was the end of it. We got along well after that. He was a helluva nice guy."[12]

Brine enjoyed the same sort of amicable workplace relationship with Maurice Ruddick. Although there was a two-decade age difference between the two men and they were never what you'd call "bosom pals," they had mutual respect. They looked out for each other. Coal miners did that—they tended to have a buddy. "You watch my back, I'll watch yours" was the informal guiding principle.

In a small, closely knit community such as Springhill, family ties were vitally important. When a miner went looking for an emergency buddy, he tended to gravitate to kinfolk, to a neighbour, someone who went to the same church, or someone with whom he played darts at the Miners Hall. Since neither Maurice Ruddick nor Harold Brine were "joiners"—Ruddick perhaps because of his skin colour, and Brine because he was inclined to keep to himself—that may well have been why and how they teamed up. Their relationship was as familiar as it was easy.

A coal miner's work is hard. It is dirty, and it is stressful at the best of times. Ever present in the back of every miner's mind is an awareness of the dangers of the workplace. A man could be working in the No. 2 mine at Springhill one moment, and he could be dead the next—another name chiselled into the wall of the miners monuments. That happened with such alarming regularity.

Try though they did, it was impossible for miners to forget all the hazards around them. Some men, Maurice Ruddick being one, loved to sing while he worked. All miners talked and joked around. "Pit talk" helped numb the pain and relieve the tedium. Conversations carried on deep underground could be as ribald as they were unfiltered. As Cumberland railwayman-turned-"miners' poet" Danny Boutilier put it, "You soon learn to swear, when you're working down there."[13] Like every man who ever worked in or around the mines, Boutilier developed an appreciation for the versatility of the word *fuck*. Why not? It's a linguistic Swiss Army knife. It can be a noun or a verb, or it can be tailored for use as an adjective or adverb. Any linguistic professor who eavesdropped on a typical coal miners' conversation would have gathered enough information for a scholarly paper.

Not surprisingly, the miners often teased each other. "Give as good as you get" was the unspoken rule of thumb where banter was concerned. No one was immune from the banter, certainly not Maurice Ruddick. However, no one called him the N-word, at least not to his face, not unless they were ready for a scrap; Ruddick had learned to box in his teenage years, and he could handle himself in a fight. No matter. He joined in the back-and-forth, and he tolerated it when co-workers poked fun at him over the size of his family. Whenever they did so, Ruddick was ready with a quick comeback. He'd laugh and explain that he intended to field his own baseball team. If a wag pointed out that Ruddick already had a dozen kids, he'd counter by explaining that he could have a football team or a choir if he and "the missus" kept having babies.

WORKING IN A COAL mine is a great equalizer, racially, socially, and in just about every other way. As one miner put it, "No individual works in a coal mine. You're all together down there."[14]

The miners' comradeship and shared sense of purpose ran deep. However, racism was as constant in a coal mine as the workplace hazards. The same held true on the streets of Springhill. Some white folks avoided talking with the town's Black citizens. Others refused to do so at all. Then there was the behaviour of the white clerks in Stedmans, the five-and-dime store on Main Street. Valerie (Ruddick) MacDonald will never forget how whenever she or any other young Black person came through the door, the clerks were like hawks watching them to thwart the shoplifting they assumed was about to take place. As she observed, "White kids didn't draw the same attention."[15]

When the racism that simmered below the veneer of polite social acceptance in Springhill bubbled up, it did so most often in the schoolyards or on the streets. It was in these venues that inhibitions would disappear. When that happened, the Ruddick kids and other Black youngsters would endure "tar babies" taunts. Occasionally, they would also hear the N-word being hurled at them. They knew such hate wasn't a spontaneous one-off; it came from somewhere deep inside and most likely was an echo of what came from the lips of parents.

Maurice Ruddick was certainly aware of all this antagonism,

but it wasn't what he was thinking about on the afternoon of October 23, 1958. It was shirt-sleeve weather, beautiful and sunny. Life was good. Wife Norma had come home from hospital just a few days earlier after having given birth to the couple's twelfth child—a daughter they called Katreena May. The proud new dad had a song in his heart and a smile on his lips while he did a few chores. One of his older girls made him a cup of tea and then packed his lunch box, filling it with a couple of honey sandwiches on brown bread, a thermos of hot tea and honey, and a piece of homemade cake or some cookies.

After giving Norma a kiss and ruffling the hair of a couple of the younger kids who were playing on the kitchen floor, Maurice Ruddick left for work. It was about 2:30 p.m. when the old screen door slammed shut behind him. He then crossed the front yard and set off across the adjacent weedy field on his five-minute walk to the DOSCO mine. He whistled every step of the way on this glorious sun-dappled day.

CHAPTER 6

THE GOOD DOCTOR

Thursday, October 23, 1958
2:45 P.M.

As Maurice Ruddick and the other men reported for work, in their minds this was just another working day in a job in which no day was ever easy or worry-free. If there was any sop to be had, it was that the weather was wondrously grand, tomorrow was payday, and it was the opening day of deer hunting season. All were reasons to feel good, unless you were a deer.

As the whistle was sounding to mark the afternoon shift change at the mine, a half-mile to the northeast Dr. R. Arnold Burden was in the middle of another hectic day. It was "situation normal" at the Springhill Medical Centre, above Wardrope's pharmacy on Main Street. Six days a week, from 8:00 a.m. until 10:00 or 11:00 p.m., that two-storey red-brick building was a beehive of activity. Burden and three colleagues who staffed the clinic—Drs. Carson Murray, J. Ralston Ryan, and DeWenten Fisher—performed surgeries over at All Saints' Cottage Hospital on weekday mornings. Whatever a patient's medical ailment was, the doctors and staff at the medical centre would treat it—everything other than brain surgeries and lung problems, which were referred to the Victoria General Hospital in Halifax, 110 miles away. If Springhill's doctors had a speciality, it was orthopaedic surgery.

The men who worked in the Springhill mine—like coal miners everywhere—suffered no end of broken bones. As Cecil Colwell, who spent thirty-three years working as a miner, told a CBC television reporter in 1985, "I had my ribs broke. My eye was injured. I've been squeezed through the neck. I had my ribs and collarbone broke. My knee was hurt. I had all the toes on my right foot broken. I injured my wrist, and the arteries got cut."

Colwell seemed just to be getting started when the interview ended. The checklist of injuries he suffered happened so often they were routine occupational hazards, part of the job. There was a good reason the DOSCO mine didn't have one of those signs you sometimes see posted outside industries that trumpet the number of "accident-free days" in the workplace. This reality would have been problematic if not for one of the benefits that was available to DOSCO employees and their families.

Prior to 1966, when government-funded Medicare became a reality in Canada, most health care was privately delivered and funded. In Springhill, DOSCO maintained a payroll deduction system called the mine "check off," which allowed miners to buy family medical coverage for fifty cents a week. "The pre-paid medical plan was ahead of its time, except in one respect," Arnold Burden once noted. "If the miner was out of work due to injury, his family was still entitled to free medical care even though he was no longer paying in. The doctor wouldn't get paid until the man was back on the job."[1]

When a patient wasn't a miner or one of a miner's family members and couldn't pay for needed medical care, the usual practice was for the doctor to add the cost to the patient's account. This happened a lot since so much of the traditional employment in Canada's Maritime provinces was tied to the land or sea. Life slowed in winter, and that meant a doctor had to wait for payment on any overdue accounts and hope they were paid in full as soon as the weather warmed up and outside work resumed.

Burden never charged interest on money he was owed, and so he frequently received goods or produce as a thank you. He loved to tell the story of how after graduating from medical school in 1952, he'd worked for a few years as a country doctor in Prince Edward Island. The lobster fishery there was big, and many fishermen were among Burden's patients. Their work being seasonal, they sometimes couldn't pay their medical bills on time. One day, when a stocky fisherman in a plaid lumberjack jacket asked him how much interest he owed on his overdue medical account, Burden informed him that he owed only the principal. Hearing this, the man asked, "Do you like lobsters?"

The conversation ended when Burden replied, "You bet!" With that, the fisherman nodded and departed. When he next appeared at the Burdens' door, it was to deliver 100 pounds of fresh lobster. "[My wife] Helen and I stayed up a good part of the night cooking lobsters, and most of them went into the freezer," Burden would laughingly recall.[2]

It was this kind of understanding and flexibility in his dealings—which was far more common in the 1950s than it is today—that people remembered about Dr. Arnold Burden. He was known to be a straight shooter, a man who displayed the kind of common sense and down-home bedside manner that earned respect. These qualities characterized so many country and small-town doctors of the day. Springhillers said of Burden that he was "small in stature but big in life."

Burden was nothing if not pragmatic. While he was always empathetic, he could also be blunt. A prime example of that occurred in his later years, while he was tending to the medical needs of inmates at the medium-security prison that opened in Springhill in 1967. One day, Burden was called upon to suture the wrist of an inmate who had failed in a suicide attempt. As the good doctor set to work with his needle, the man demanded, "Doc, aren't you going to give me anything for the pain?"

"Why?" Burden asked. "You didn't have anything when you slashed your wrists."[3]

IN PRE-MEDICARE DAYS—that is, long before so many people sought out their medical advice on the Internet or heeded the directions given by the self-appointed "experts" who appear online or on television—general practitioners (GPs) were the backbone of Canada's health care system (just as they still are today). They were jacks of all trades, masters of many. If the concept of work–life balance even existed for them, it was seldom a topic of conversation in the profession. In Springhill, as was the case in so many other small towns, GPs were on call 24/7. They made home visits at any hour of the day or night, seven days a week—for which they charged two dollars per visit in the 1950s—and since there was no ambulance service in thinly populated areas, doctors often used their own cars to transport patients to the medical centre or the hospital. Recalled Arnold Burden's son Bill, "For many years, Dad's winter go-to [vehicle] was an unheated World War II–vintage Jeep that was held together with spit and electrical tape."[4]

Burden had vivid memories of how when he'd worked as a country doctor in Prince Edward Island "it was commonplace to make house calls in a car or by horse and sleigh over 'winter roads'—that is, through orchards, across fields, and alongside the woods, leaving the snowplows far behind."[5] On one occasion, when the spring melt had made off-roading impossible, Burden drove a circuitous 150-mile route on back roads to treat a child who had a fever of 106 degrees Fahrenheit.

Then, as he did after returning to Springhill in 1957 to take a job at the medical centre, Burden more than earned his money. His days were long, and his responsibilities were onerous. Writing in her memoir, singer Anne Murray—Springhill's most famous native daughter—recalled a typical workday for her father, Dr. Carson Murray, who was Burden's colleague and friend. *Exhausting* is the descriptor that immediately springs to mind.

Anne Murray noted that her father's workdays began before 7:00 a.m. An hour or so later, he'd be busy performing or assisting with morning surgeries at the medical centre. This was followed by patient rounds at All Saints' Cottage Hospital, the rambling Victorian-era structure that served as the town's infirmary. After that, Murray did two daily sessions—one in the afternoon and another in the evening—treating patients who lined up on a first-come, first-served basis at the medical centre, and then he did a second round of patient visits each night at the hospital. Most nights, Murray got home around ten o'clock. This gave him just a few precious minutes with his wife and any of his kids who were still awake. Despite what you'd expect, the doctor's day didn't end there. Before going to sleep, he often had homework to do. "Dad kept up professionally by reading exhaustively, often late into the night, from medical journals he kept stacked beside the bed," Anne Murray recalled.[6]

That was pretty much how every working day went for the four doctors who staffed the Springhill Medical Centre in 1958. Small wonder that at that time, more than is the case today, physicians were respected—and even revered and beloved—members of their communities. For them, a medical career was more than a job—it was a way of life. That was true for Drs. Carson Murray, DeWenten Fisher, Ralston Ryan, and Arnold Burden.

All four men were cut from the same bolt of cloth. All hailed from the same general area of western Nova Scotia—two of them from Cumberland County and two from neighbouring Colchester. All were graduates of the Dalhousie University Medical School in

Halifax. And three of the four had served in the Royal Canadian Army Medical Corps during WWII. Ralston Ryan was the sole exception; he had been deemed medically unfit for military service owing to a heart murmur.

Fifty-year-old James Carson Murray, the doyen of Springhill's doctors, was born in 1908 in the Colchester County village of Tatamagouche, 45 miles east of Springhill. As his daughter Anne remembers, when her dad arrived in Springhill in 1934, he was "a handsome young Dalhousie medical school grad who had just completed a year of surgical training at St. Luke's Hospital in Cleveland (later the Cleveland Clinic) and a year's practice with his father, a country doctor . . . in Tatamagouche."[7] Although he'd been raised as a Presbyterian, Carson Murray fell in love with and in 1937 married a nurse at All Saints' Cottage Hospital named Marion Burke, who was a Roman Catholic. This twenty-four-year-old coal miner's daughter from the nearby town of Joggins was ever personable, gracious, and outgoing. In those regards, she was the perfect match for the soft-spoken, introspective Dr. Murray. Despite their different religious upbringings and the initial misgivings of parents on both sides, the couple made their marriage work for forty-three eventful, but happy, years.[8] Together, Marion and Carson Murray raised six children and stood as pillars of the community who involved themselves in almost every aspect of town life.

DeWenten H. Fisher, who in 1958 was thirty-four years old, hailed from Great Village, a Colchester County hamlet 40 miles southeast of Springhill. The son of a carpenter, he was the only one of the four children in his family to attend university. He dedicated himself to the practice of medicine, so much so that he never married. It was typical of the man that when he received a medical emergency phone call at 2:00 a.m. one snowy winter night, he got out of bed and set to work shovelling his car out of the driveway. When a follow-up call came to let Fisher know the ailing patient had died, Fisher made the house call anyway. Asked later why he'd bothered to do so when a blizzard was still raging, Fisher's brother Russell remembers DeWenten's reply, which was typical of his approach to his work. "The family expected me to come. I had to fill out the death certificate," he said.[9]

J. Ralston Ryan, who was born August 22, 1914, a few weeks after the outbreak of WWI, was the older son of Springhill businessman Thomas B. Ryan and his wife, Ethel (Hall). After graduating from Dalhousie medical school in 1939, Ryan returned to his hometown at age twenty-four to help care for his aging parents. In addition to practising medicine, he had a head for business. Following in his father's example, Ryan profited as a result of his involvement in various local businesses, pharmacies being among them. Then, in 1949, he and Carson Murray teamed up to open the Springhill Medical Centre.

Like Ralston Ryan, R. Arnold Burden was a Springhill native. And like Ryan and his other two colleagues, Arnold Burden was passionately devoted to his profession, his patients, and his community. However, there was something else that was unique about Burden, and it created a special bond with his patients who were miners: Burden himself had worked in the DOSCO mine for two summers when he was at medical school. He never forgot how the miners often went out of their way to help him with the demanding work he was doing. "You watch those hands, kid, you're gonna be a doctor," they told him. As family members noted in his 2018 obituary, "Arnold never forgot his friends who toiled [there] and how they had helped to ensure he was able to obtain his medical degree. Small wonder that he cherished this bond."

Arnold Burden was born April 28, 1922, into a mining family. His grandfather had managed two Springhill collieries in the 1920s, and his father, George, worked as a DOSCO machinist. Burden's mother, Mary (née O'Rourke), died in 1926 when she was just thirty-two, and so young Arnold and his baby sister, Audrey, were raised by their father. They came of age during the lean years in the Great Depression of the 1930s. Money was tight, and life was anything but easy, especially in a one-industry town like Springhill.

Dr. Arnold Burden was a Springhill native who'd worked as a miner during his days as a medical student.

(Courtesy of the Burden family)

As a young man, Burden was diminutive, scarecrow thin, and bespectacled. He was also whip smart, a voracious reader, and a gifted student who excelled academically. In his teenage years, incongruously perhaps, he became an avid hunter, fisherman, and backpacker. He was also artistically inclined and loved to sketch and paint, carve wood, and take photographs. When WWII broke out in Europe, Burden was eager to join the army and serve overseas, where the action was happening. But in the spring of 1940, he was a scrawny seventeen-year-old who tipped the Toledo scales at just 110 pounds. There wasn't much to him. No matter. Burden was eager to enlist anyway. The medical officer at the recruiting depot in Amherst, Nova Scotia, took one look at the skinny teenager standing before him in his underwear and barked, "Get the hell out of here, and don't come back without a birth certificate!"[10]

Undeterred, Burden persisted in his efforts to enlist. Thinking his luck might change if he tried to sign up in a different city and with a different branch of the military, Burden travelled to Moncton. There he tried to join the air force. Not surprisingly, he had no more luck than he'd had in Amherst. Having endured a second rejection, he returned home to Springhill, where he bided his time working at a grocery store. He was too slight to work in the coal mine; a man had to weigh at least 120 pounds to do that.

By early January 1941, Burden was eighteen and still determined to serve King and country. Intent on finally enlisting, he travelled by train to Halifax. At the time, this was the farthest away from Springhill he'd ever been in his life. His adventure continued, for this time out he was accepted into the army. The four years he would spend in the Royal Canadian Army Medical Corps (RCAMC) forever changed him and put him on a career path to become a doctor.

The No. 7 Canadian General Hospital, with which Arnold Burden served, was the first Canadian medical unit to land on Juno Beach after the Allies' June 6, 1944, D-Day landings. For the next year, the men and women of the No. 7 Hospital cared for the wounded as the First Canadian Army battled the German army, clawing its way across Belgium and Holland and into Germany. In early 1945, Arnold Burden was there when the Canadian Army liberated the Sandbostel concentration camp, near the city of Hamburg. Burden was sickened, angered, and revulsed by the sights and stomach-turning smells of the railway cars that brimmed with corpses. And he felt haunted by his encounters with emaciated inmates whose skin was sallow and covered with sores. Not surprisingly, the horrors he'd witnessed during the war affected him deeply. When he returned home to Springhill, he did so with an ambition to become a doctor.

Taking advantage of government programs that provided veterans with free university tuition and a small monthly stipend, in 1946 Burden enrolled at Dalhousie medical school. While it wasn't cheap to live in Halifax, he managed to make do, and in August of 1948, he took the plunge and got married. His bride, Springhill-born Helen Dewar, was six years younger, and so they hadn't met or known each other when they were growing up in Springhill. Regardless, when a mutual friend introduced them, it was love at first sight. Through good times and bad, the couple would remain happily wed for seventy years and would raise four sons.

To help pay his way through medical school, Burden spent those two summers working in the No. 1 and No. 4 DOSCO collieries at Springhill; he'd grown and put on enough weight in recent years that he'd met—albeit barely—that 120-pound minimum weight requirement to work in the mine. Doing so engendered in him a deep empathy for the men who mined coal for a living. Burden learned first-hand how hard and uncertain a miner's life was.

Working underground took some getting used to. The only light that pierced the perpetual darkness came from the bobbing caps of the miners as they went about their tasks. Then there was a peculiar odour that Burden described as "mine smell." It was "not unpleasant, but awfully distinctive. It was the odour of air that had been pumped down into the mine, circulating three miles through the maze of underground tunnels and mixing with the sweat of hundreds of toiling bodies."[11] The temperature this far underground was surprisingly constant—usually about 80 degrees Fahrenheit—pretty much year-round.

The mine's workaday sights and smells were unforgettable. No one who experienced them ever forgot them. Nor did anyone forget the mine's soundtrack. It was a cacophony made up of the distant whirring of the huge ventilation fans that circulated fresh air through the mine's labyrinth of tunnels, shafts, and crosscuts; the periodic rattling of coal-rakes that moved coal down the slopes; the chipping of the axes, picks, and shovels the miners used to dislodge coal from the eight-foot-high faces on each level of the mine; and, of course, the chatter, grunting, cussing, and singing of the miners as they went about their daily labours.

Other constants were the dangers posed by noxious gases and coal dust and the unease that came from being continuously observed by a mischief of hungry rats. "Each noon hour as I unpacked my sandwiches, I could hear the loose coal or stone rustling. Since it was pitch black, except where the light on my forehead was pointing, I couldn't see them until I pointed in their direction," Burden would recall. "There they were, six or seven feet away, grey, and about ten inches long. . . . Some would scurry away and hide, while others just stood there fixing their beady eyes on me."[12]

The darkness, the pungent smells, and the presence of voracious rodents all served as vivid reminders of what an inhospitable, alien environment a coal mine is. It's one in which a person can suffer horrible injury or death in an instant and without warning. Although the miners with whom Burden worked did their best to coddle him, that workaday reality was something he came to know all too well even in the relatively brief time he worked as a miner. He had several close calls with death.

On one occasion, he was underground when one of the mini-bumps that routinely rattled the mine occurred. Some coal fell from the roof of the level where he was working, and for about ten minutes a thick cloud of coal dust fouled the air. Burden was shaken but unharmed. Another time, when a coal face collapsed without warning, a falling chunk of the mine's roof struck him on the back. As he scrambled to find shelter between the packs, loose rocks and coal debris buried one of Burden's legs. Fortunately for him, he sensed what might happen and was able to jerk his leg free an instant before more of the roof came crashing down. Although Burden was bloodied and bruised, he escaped serious harm. His run of good luck continued a week later when a telephone pole–sized chunk of wood suddenly fell out of the roof of the mine and whacked him on the head. Had it not been for his safety hat, that might well have been the end of medical student R. Arnold Burden. As it was, the force of the blow knocked him flat and pinned his shoulders. With no one else close at hand or within earshot, Burden was on his own. It took him no small amount of frantic effort to free himself.

You might think such experiences would have frightened Burden to the point he'd never go down into a mine again. Not so. The doctor was living and working in PEI in 1956 when the fiery explosion that killed thirty-nine men rocked the No. 4 mine. Sensing there'd be miners in dire need of medical attention, Burden packed his car full of drugs and bandages and raced to Springhill. There he treated burn victims before volunteering to go into the mine with rescuers. At one point, while he was crawling on hands and knees to rescue some injured men who'd been overcome by firedamp, Burden himself passed out. Fortunately, he came to after collapsing to the mine floor, where the air was still breathable. Undeterred by this near-death experience, Burden persisted in his volunteer rescue efforts both that day and the next. While doing so, he treated many of the eighty-eight men who were rescued from the mine.

Arnold Burden certainly knew all there was to know about the on-the-job dangers that coal miners faced every day. Like the miners who worked in the Springhill mines, he also understood it was the luck of the draw who lived and who died when bad things happened—as they were fated to do—on the afternoon of October 23, 1958.

All Saints' Cottage Hospital, the town hospital in Springhill, was a rambling Victorian frame building with beds for fifty-six patients. (Courtesy of Billy Kempt)

CHAPTER 7

GOING TO WORK

THURSDAY, OCTOBER 23, 1958

3:00 P.M.

THE WELL-CHOREOGRAPHED ROUTINE WAS AS CONSTANT as the passing of time in Springhill. It had rolled out pretty much the same way for the eighty-five years of commercial coal mining in the town. And so it was that a few minutes before three o'clock on this sunny afternoon, Harold Brine, Maurice Ruddick, Tommie Tabor, and the other 174 men working the afternoon shift at DOSCO's No. 2 colliery gathered at the corrugated tin–sheathed wash house.

It was here in this drafty barn-like building that the miners donned their work clothes. It was also here that on a normal day, they'd shower and change back into their street clothes at the end of their eight-hour shift. There were no lockers in the wash house. Instead, each man accessed a storage bucket suspended from the ceiling by a system of ropes and pulleys. After lowering his bucket to the floor, a miner retrieved and donned his work clothes (which as often as not were sweat-stained, crusty, and malodorous), and then raised the bucket containing his streetwear back into the rafters. There it hung until he returned at the end of the shift.

A coal miner's simple kit was well suited to the harsh conditions of underground work. Nowadays, common sense and government health and safety regulations would dictate that the men be equipped with and wear safety gear. Not so in 1958. The world was a very different place. There doubtless were many doctors at that time who would have sworn that smoking wasn't harmful to your health and provided ample testimony to that notion. As late as 1946, the American tobacco company R.J. Reynolds had mounted an ad campaign featuring the slogan "More doctors smoke Camels than any other cigarette."[1]

And so it was that for whatever reason—a lack of awareness of the true dangers of the environmental hazards that existed in a coal mine, simple indifference to those hazards, or perhaps because macho dictates insisted that "real men git 'er done"—coal miners wore no protective gear. None. They had no eye goggles, no earplugs, and no particle masks to filter the coal dust–laden air.

Mind you, the No. 2 colliery at Springhill was as well ventilated as it was possible to make it. Despite this, the dangers of poisonous gases were ever-present in any coal mine. So too were those posed by the particles in the air.

Mining invariably creates clouds of dust that cling to and blacken a miner's clothes, skin, and tools. The dust also seeps into the lungs. It accumulates there and pits the living tissues. The by-product of this damage is scarring that hardens the lungs, reduces their efficiency and elasticity, and makes it increasingly difficult to breathe. This in turn leads to a drop in oxygen circulating in the blood. Doctors call this incurable condition a type of pneumoconiosis; miners referred to it as black lung disease. Among veteran coal miners, it was as common as callused hands; however, it was far more problematic and deadly. Black lung disease kills quietly and slowly. Herb Pepperdine, like so many of his long-time co-workers, developed black lung disease during his eighteen years working underground. "I think we all got it," he said. "Nothing you can do. Gets you in the end, I guess. [And when] it's getting to you, you can hardly breathe."[2]

Leaving aside heart conditions, cancers, and the myriad other occupational health hazards associated with coal mining, the numbers tell us that in the United States alone, black lung disease killed more than seventy-six thousand miners in the half-century between the late 1960s and 2014.[3] That's far more than the number who died in cave-ins, explosions, and all the other underground disasters that plague coal mines. However, until the 1950s, no one really knew or cared much about black lung disease. The great irony in this situation was that in the latter decades of the last century, its pervasiveness as an occupational hazard among coal miners was due in no small measure to the priorities of the leaders of the miners' union, which was American based.

Springhill was a stalwart union town with a long, robust history of collective action. In 1879, when a coal miner's take-home pay was only about eighty cents per day,[4] a decision by the mine's owners to impose a ten percent wage cut prompted a strike and the formation of the Provincial Workmen's Association (PWA), which was North America's first coal miners' union. Then in 1908, the United Mine Workers of America (UMWA) muscled in and made a determined bid to supplant the PWA. The two rival unions competed for members until 1917, when they amalgamated. And the following year, Local 4514 of the UMWA began representing Springhill miners.

Integral to any mention of black lung disease in the Springhill coal mine or elsewhere at that time was the DOSCO pay system, which compensated miners for the tonnages of coal they produced each shift. The more coal the men extracted from the coal seams, the more they earned in addition to their base hourly wage. In that regard, the introduction of new, more efficient mining drills in those mines across North America where rock-hard anthracite coal was mined was money in the pockets of the miners and their union. However, the proverbial sword was double-edged: with increased production and those much-desired bonuses came more dust, greater health risks, and a rising incidence of black lung disease among coal miners.

It's unclear whether John L. Lewis, the long-time, all-powerful Washington-based president of the UMWA,[5] was aware of any of this, much less cared about it. He had other, more immediate and pressing concerns. Lewis's key priority was boosting union members' paycheques (and union dues). Part and parcel of that bread-and-butter concern was his desire to strengthen the UMWA's welfare and retirement fund. Higher outputs of coal were pivotal to achieving both those goals.

For better or worse, anthracite is harder than the bituminous coal that was found at Springhill. Miners who worked in the mines here didn't use electric drills or dynamite; the danger of sparks igniting the ever-present methane gas was too great for that. In the early years, Springhill miners used picks, hammers, and shovels to extract coal. By the early twentieth century, they were doing some tunnelling with help from compressed air–powered chippers—which were like the jackhammers you see roadwork crews using today. However, for the most part in their work, the men continued to rely on old-fashioned, labour-intensive muscle power. Doing so created less dust than fouled the air in mines where drills and explosives were commonly used, yet black lung disease remained as big an occupational hazard in Springhill as it was elsewhere.

All questions of workplace safety and union priorities aside, most Springhill miners wore whatever workplace attire was practical and comfortable. And so most men were clad in pants and a T-shirt that could easily be shed if necessary. It doesn't get any more basic than that. "You'd work up a good sweat when it got hot down there," Harold Brine recalled. "I'd often take off my shirt and go bareback, work just in my pants. A lot of the boys did the same thing."[6]

Many of the savvy old-timers had a slightly different way to keep cool. They wore sleeveless wool pullovers that wicked away and absorbed sweat. Other than that, the outfit the miners wore was pretty much standard: a pair of steel-toed work boots, heavy leather gloves, and a belt with a pouch for the rechargeable battery each man carried. That vital bit of technology—which was the size and weight of a hardcover copy of *War and Peace*—powered a lamp that clipped into a bracket above the brim of the man's safety helmet, or "cap," as the miners called it. These lamps, which were a brainchild of the renowned American inventor Thomas Edison,[7] were about as bright as the appliance bulb you have in your refrigerator, and they shone for as long as twelve hours. The miner's headlamp was his lifeline in every sense of the word. Without a lamp, he was blind and couldn't see his hand when it touched his nose.

After changing into their work clothes, the Springhill miners walked next door to the adjacent red-brick building they called the lamp cabin. Here, row upon row of the batteries that powered the miners' lamps were plugged into the huge electrical panel that recharged them. At the start of shift, the men lined up to get their lamps; afterward, they lined up again to return them and retrieve the octagonal-shaped, toonie-sized brass tag that every miner carried. This identification (ID) tag bore a stamped number two—designating the DOSCO No. 2 mine—and the miner's ID number. The latter was as much a part his workplace identity as his name; Harold Brine was No. 660, Maurice Ruddick No. 624, Tommie Tabor No. 872.

The man in charge of the ID tags was Harry Weatherbee, a reed-thin, bespectacled Springhill native who'd worked at the mine for forty-two of his fifty-six years. He and his assistant, Alf Cox, managed the tags and maintained the miners' lamps, ensuring that each one was fully charged and in working order.

Although there was little time for small talk during a shift change, Weatherbee and Cox knew every miner by name and number. After collecting a man's ID tag at the start of a shift, the lamp men handed over a lamp and a battery. They then hung the miner's ID tag on a hook on the big wall board that showed at a glance who was working that day and how many men were underground. If a miner failed to reappear to turn in his lamp at the end of his shift, Weatherbee and Cox knew it.

The ID tags were vitally important to the mine's operation. It was a big deal if a man forgot his tag or lost it; the general rule was "No tag, no battery or lamp." On October 23, veteran miner Percy Rector didn't have his ID tag, No. 1202.[8] He'd forgotten it at home. This fifty-five-year-old Springhill native, a big, good-natured man who laughed easily and was everybody's friend, was a "lifer." He'd worked in the mine since 1921, when he was a fuzzy-cheeked eighteen-year-old. Mining was the only job Rector had ever known, and in his thirty-seven years working underground, he'd seen and done it all. Nothing much fazed him; he could deal with whatever problems life or his work threw at him. Despite his easygoing nature, not having his ID tag was an issue for him today.

On this Thursday afternoon, as he sometimes did before punching in, Rector had stopped in at the Miners Hall on Main Street with his pal and brother-in-law, Leon Melanson. The two had played darts and enjoyed a few bottles of soft drink. They relaxed and had a grand time. Maybe that was why Rector forgot his ID tag. Who knows? However, none of that mattered to Harry Weatherbee. All he cared about was that if a miner didn't have his ID tag, he didn't get a lamp; that was company policy. If Rector wanted to work today, he was obliged to go to mine manager George Calder's office, explain his situation, and fill out a form that would net him the requisition slip he needed to check out a lamp and battery without his ID tag.

Springhill coal miners—like all miners—were superstitious; they regarded any number of occurrences as bad luck. For example, it was taboo for a woman to enter a mine. Some miners were deathly afraid of being the first to chip away at the coal face when a new shift began. Another no-no was that if a man emptied his water canteen while he was still on the surface, it was bad luck for him to refill it *before* going down into the mine. Surprisingly, forgetting your ID tag evidently wasn't considered to be a portent. Or if it was, apart from a bit of good-natured ribbing that Percy Rector took for his absent-mindedness, no one mentioned it to him on this day, and he wasn't thinking about it. After belatedly collecting his lamp and battery at the lamp cabin, Rector rushed to catch up to his workmates. By now, they were gathered over near the pithead, about 75 yards distant across a weedy field.

When the weather was warm, miners often sat around near the hoist shed killing time while they waited for their shift to begin. (Nova Scotia Archives)

AS THE LAST few minutes before three o'clock ticked away, the 174 men who were on the afternoon shift in the mine milled around the hoist shed. Today, as on any day that the weather was warm and the sky was clear, the men would bask in the sunshine while they talked and watched the cable hoist do its work. The huge electric motor whirred and groaned as it pulled on and let out the inch-thick cables that moved trolley cars up and down the mine's 30-degree main incline. These low flatbed cars looked like grimy versions of roller coaster trains, except they were counterbalanced in pairs—one was always descending while its partner was ascending on the double-tracked rail system. With eight linked trolleys on the go at once, and with ten men riding in each train, eighty miners could ride up or down at the same time. The shift change went quickly and smoothly most days.

The seams in the upper levels of DOSCO's No. 2 mine at Springhill had been exhausted for many years. As a result, the coal being extracted in October 1958 came from coal faces much deeper—as far down as the 13,800-foot level. The No. 2 Springhill mine was deep. It was the deepest in North America; some said it was the deepest in the world. "When I went [down] in No. 2, that's the farthest I was ever away from home in my life," veteran miner Herb Pepperdine liked to say.[9] He wasn't alone in that regard.

The 2.5-mile track distance between the pithead and the No. 2 mine's nethermost depths was too great for efficient operation of a single cable hoist, and so there was a second unit deep underground. Located at the 7,800-foot level, it was the hub of the transfer station (akin to the busy Bloor–Yonge subway stop in Toronto). After exiting the cars that carried them on the first leg of their downward journey, miners walked along the level connecting the mine's main slope to the "back slope." Here, a second hoist system moved the trolley cars up and down along the tracks that connected the transfer point to the deepest recesses of the mine and the faces where coal was being extracted.

Many of the smokers who were waiting at the pithead on the afternoon of October 23 hurriedly puffed on the last cigarettes they'd enjoy for eight hours. Smoking and open flames were strictly forbidden in the mine. Those miners who were in the habit of lighting up—and that was most of the men—had no choice but to chew tobacco or do without their nicotine fix during a shift. The tobacco chewing habit was rampant, both because it satisfied a man's cravings and because some miners claimed it helped clear coal dust out of the mouth and throat.

While they waited for the three o'clock start of their shift, some miners killed time by filling their metal water canteens—but only if the cans were empty and hadn't been filled and emptied earlier. Remember: refilling them then was regarded as bad luck. Others talked about family matters, or else they marvelled at the day's fine weather while grousing about having to go to work when they could have been doing work at home—as Harold Brine would have been doing—or maybe getting ready to go deer hunting; the annual season was due to open the next day. Then there were always miners who talked about sports.

As Tommie Tabor would have told you, Springhill was a baseball town; it wouldn't be a stretch to say it was baseball crazy. However, by this late in October, the local sandlot season was long over, and in the big leagues, the World Series had wrapped up two weeks earlier. The New York Yankees had bettered the Milwaukee Braves in a thrilling, hard-fought seven-game final that still had ball fans talking. Now that the weather was starting to cool off, the conversation naturally turned to hockey. The puck had dropped a couple of weeks earlier on the NHL's 1958–59 season. Most Springhill hockey fans cheered for the Montreal Canadiens and their aging star Maurice "the Rocket" Richard, but the Boston Bruins and the Toronto Maple Leafs also had plenty of supporters in Springhill.

Conversations among the men of the afternoon shift about hockey and everything else were cut short, and the smokers took one last deep, lung-filling drag before they flicked away their cigarette butts. A man-rake burst out of the darkness of the main slope and came to a jolting halt inside the hoist shed. When it did so, the first of the miners who'd worked the morning shift came spilling out, their muscles aching, hands and faces blackened with coal soot, and the whites of their blinking eyes glistening beneath the brims of their grimy caps. The men made a beeline for the lamp cabin and then the wash house. Another day done. And tomorrow was payday.

Harold Brine, Maurice Ruddick, Tommie Tabor, and the rest of the men who were waiting to go below scrambled to fill the freshly vacated seats in the man-rake. On the descent, the most coveted spots were in the back row. As Harry Munroe, a Springhill miner with a talent for writing poetry (and who'd been a barefaced mine rescuer in 1956 and again in 1958) explained in one of his verses, "The top seat in the trolley was considered the best / Because while riding they could lay back and rest."[10]

Some miners liked to catnap on their downward journey; others sat quietly or chatted with their seatmates. Some liked to sing during the descent. Maurice Ruddick—"the Singing Miner"— was among them. Today, as he was wont to do, Ruddick broke into song. In his mind, there was no better way to begin a work shift. Ruddick's choice of music was usually a pop song of the day or maybe a spiritual. Sometimes it was one of his own tunes; "The Curse of the Old Number Two" or "Away Down in the Deeps" were two of his favourites. Ruddick's co-workers had heard him sing both these songs so often that they knew the words almost as well as he did. Some of them would join in as he sang. Today, as the wheels began to clickety-clack and the trolley disappeared down the main slope into the world of darkness below, the last lines of "Away Down in the Deeps" were left hanging in the air at the pithead: "If we should die in the deeps, boys, / we know where we will go, / Heaven, boys, is away way above, / and hell is right here below."[11]

96

Within moments, as the trolley began its descent of the main slope, the sunlight was just a memory. The men found themselves in the eternal darkness of this subterranean world. It was a world they all knew too well. The blackness was so all-encompassing and overwhelming that you could almost touch it. The only light, that which emanated from the lamps mounted on the crest of each man's cap, sent shadows of the trolley riders dancing along the walls of the mine shaft. At the widest point, it was about 10 feet across. "In some places the roof was quite low, supported by steel railroad rails that in their turn were supported by wooden packs on either side of the slopes," Arnold Burden wrote after taking his first trolley ride down this same slope. "Gaping hollows overhead showed where rock had broken loose and fallen onto the slope; in these sections, the roof was so high my light was only a faint beam at the top."[12]

About forty minutes after starting their descent into the mine, Maurice Ruddick, Harold Brine, and the others in their work crew arrived at the 13,000-foot level. When the man-rake came to a stop, a crew of miners piled out. The shift manager had assigned a couple of dozen men to work on the coal face there this afternoon. Some of the miners in the rake continued down to the 13,400-foot level; Tommie Tabor and the rest of his crew stayed in their seats until the last stop, the 13,800-foot level.

When the last of the miners had exited the rake, the second contingent of men who had just completed the morning shift quickly piled in, filling the seats.

IN OCTOBER OF 1958, miners in the No. 2 mine were actively working those three adjacent coal faces—one on the 13,000-foot level, a second at 13,400 feet, and a third at 13,800 feet. (A fourth level, at 14,200 feet, while still in development, served as a ventilation conduit.) The sights, smells, and sounds at these great depths were very much those of an alien world. Yet many of the men who toiled underground enjoyed their work; some even loved being down there. They relished the routine, the feeling they were doing a vital job that not everyone could do. And most of all, they loved the sense of camaraderie. Gorley Kempt certainly did.

Thirty-eight years old and a veteran of seventeen years in the mines, Kempt would have told you he was born to be a coal miner. Who knows? Maybe he was. As Kempt explained, "On top, I miss the camaraderie we have in the pit; [there's] no jealousies, no rank down there. We're dependent on each other and in a crisis a man's first impulse is to help a buddy even if it means a fifty-fifty chance of getting hurt himself."[13]

Small wonder that many a miner referred to the mine the same way sailors would speak of their ship—that is, with affection and using the *she* pronoun. "It's in the blood" was a phrase often heard when miners talked about their affinity for their work, their workmates, and the mining culture. As Maurice Ruddick's wife Norma would observe many years later, "I think all the miners [felt that way] because it was their life, eh? Mining was their life. Nothing else."[14]

Herb "Pep" Pepperdine, who was one of Ruddick's co-workers, echoed that comment. "Everybody was friends," he said. "[We had] lots of fun . . . and joked and called one another names and had nicknames. . . . Oh, I enjoyed it."[15]

Almost half of the men on the afternoon shift—Maurice Ruddick, Harold Brine, and Tommie Tabor among them—were "hands on," doing the actual mining of the coal on the coal faces. There were miners working about every 15 feet along the length of each level of the mine. About a third of the men were labourers rather than miners. Their job was delivering and positioning the timbers used to build the packs that held up the roof of the mine, positioning the conveyor pans, and doing the myriad other support jobs that needed doing. The remaining members of the crew were preoccupied with the unending work of making sure the mined coal moved along the level and out to the slope, where it was loaded onto the rakes that carried it to the pithead.

There, in the light of day, after having lain buried for as long as 400 million years, the coal glistened like the Industrial Age black gold that it was. Understandably so, for coal is the quintessential fossil fuel. The energy stored in coal originally came from sunlight that eons ago nurtured the ferns and other green plants, which then died, decomposed, and subsequently came under incredible pressures deep below the surface of the earth.

Those were the sights, sounds, and smells you would have experienced if you'd been at the pithead or if you had ventured down into the No. 2 mine at Springhill on the afternoon of October 23, 1958. To the men who were labouring there, it was just another day at work. Or so it seemed.

CHAPTER 8

A HARBINGER OF THINGS TO COME

———

THURSDAY, OCTOBER 23, 1958
6:00 P.M.

THE DAY'S LAST RAYS OF SUNSHINE WERE CASTING LONG, slanting shadows across the Cumberland landscape on this splendid autumn evening. The streets of Springhill, which had been bustling with vehicular and pedestrian traffic just a few hours earlier, were now relatively quiet. At this hour on a weekday, the shops and businesses along the main drag had closed for the day. Schools had been out for two hours. Despite the warmth in the air, the parks and playgrounds were all but empty. Most Springhill residents, young and old alike, were at home at this hour. It was suppertime. And home is where on an autumn evening most people stayed after eating. Adults did the dishes, read the newspaper, listened to the radio, played cards, or maybe sat on the porch talking and smoking if the weather was nice. Kids did their homework or played in the yard until it got dark. For those locals who were fortunate enough to have a television, it was shaping up to be a promising night for viewing. Those who didn't have a television invented an excuse to visit neighbours who did. Tonight, they had a valid one.

Four young people from Springhill High School were scheduled to sing on the musical variety program *High Society*, which was due to air locally at seven o'clock by television station CKCW, the CBC affiliate in nearby Moncton, New Brunswick. Right after that, it would be time for *I Love Lucy* and *Don Messer's Jubilee*. Both shows were hugely popular.

For Springhill residents who had rooftop television antennas, the signal for Channel 2 came in loud and clear. Those who relied on rabbit ears, depending on weather conditions, sometimes had to raise or lower the angle of the metal arms in order to capture the elusive signal. Fortunately, on a cloudless day like today, reception was good; there was none of the "snow" interference that sometimes made it difficult to watch a program.

While he might have wanted to stay home, there would be no television viewing for Dr. Arnold Burden on this night. The unseasonable warmth of the afternoon was fading quickly as he motored to the Springhill Medical Centre. It was Burden's turn to be one of the doctors on duty for evening office hours; doing so was a part of his job that he didn't mind. As he drove along McFarlane Street, the doctor was in a good mood. As usual. He had just finished eating supper with his wife, Helen, and their boys. The Burdens lived in a big old white-frame house that was just a couple of blocks from the medical centre, but Arnold Burden always drove his old Jeep on the commute to and from work. He needed his vehicle to visit patients in their homes and to drop by the hospital for late rounds. That was his usual evening routine.

They say having a routine is key to staying organized when you are crazy busy. Sound advice, that. And having a routine can also be an effective coping mechanism for people who find themselves in a stressful or difficult situation. This may help explain the workplace behaviour of miners on the 3:00 to 11:00 p.m. shift at Springhill's

No. 2 colliery. At six o'clock, barely three hours into their eight hours of work, the men were in the habit of taking twenty minutes off to eat a quick meal; this was at the same time their families at home were sitting down to eat supper. And so it was on this Thursday evening down in the mine. The sameness of the routine was reassuring, and it was easily taken for granted; no one gave it a second thought. However, the clock was ticking down on the last few minutes of normalcy for the residents of Springhill.

Meanwhile, over at the medical centre, Dr. Burden was doing what he did most evenings when he was on duty. After thumbing through some patient medical charts, he was ready to see the people already lined up in the waiting room. It promised to be another busy evening.

Things were shaping up to be much quieter out at the Brine family's residence on Mountain Road, where Joan Brine was feeding her two-year-old daughter, Bonnie. The pair were seated in the kitchen, the only room in the still-under-construction home that Harold Brine had finished so far. Rome wasn't built in a day; neither is a house.

A mile to the east of the Brines' place, at suppertime the kitchen in the Ruddick house on lower Herrett Road was thrumming like a bag of bees. It was always that way at mealtime. The family would gather around the two old wooden tables that sat in the middle of the kitchen floor. Norma Ruddick typically baked as many as ten loaves of bread each day for her family, and she would make a stew or a large pot of skim milk and peas with bread and molasses. The latter was a Ruddick family favourite recipe that wasn't found in the Purity Flour Mills' promotional cookbooks that were standard references in the kitchens of many Canadian homes of the day. However, a week after having given birth for the twelfth time, Norma Ruddick wasn't preparing much of anything. She was feeling poorly, and so she was taking the evening off. Although she was still only thirty-three, childbirth had never gotten any easier for her.

With the Ruddicks' three eldest daughters in charge of the cooking tonight, Norma was more than happy to sit with seven of her younger children. While waiting to be fed, they stared at the empty chair at the head of the bigger of the two tables. This was Maurice's spot. He presided over family meals when he was home. On work nights, he was conspicuous in his absence.

With a dozen children at the table, as this photo by *Chronicle-Herald* photographer Maurice Crosby shows, mealtimes were always busy in the kitchen of the Ruddicks' home. (Nova Scotia Archives)

At about the same moment his wife and kids were having supper, Maurice Ruddick was thinking of home and family. He, Harold Brine, Tommie Tabor, and most of their co-workers were sitting down on the first of the two twenty-minute break periods that punctuated each eight-hour shift at the DOSCO mine. Perched there on the floor of the mine, in the pitch black, deep underground, the men opened their company-issued sheet-metal lunch boxes, which they were required to purchase from DOSCO.

The metal lunch boxes carried by miners were company-made and were sold by DOSCO. (Ken Cuthbertson)

Most of the men devoured a sandwich or two and then chased the food with mouthfuls of lukewarm water they sipped from their canteens, or with coffee or tea quaffed from their Thermos bottles. With coal dust everywhere and their hands and faces soot-blackened, the men handled their sandwiches carefully—that is, only by the crusts. After eating only the middle part of the bread, they'd toss the soiled crusts over a shoulder and into the darkness where the mine waste was piled—into the "gob," as it was called. Small wonder that rats were so plentiful and plump in the No. 2 mine. They were also canny.

Miner lore has it that rats, like many other animals, can sense that an earthquake or other seismic disturbance is coming, and they make themselves scarce. There's a growing body of evidence suggesting that this idea may be true. The earliest reference we have about unusual animal behaviour prior to an earthquake comes from ancient Greece. In 373 BC, the rats, weasels, snakes, and centipedes all reportedly disappeared several days before a destructive earthquake occurred. And there is a huge volume of other anecdotal evidence about various critters behaving strangely anywhere from weeks to seconds before an earthquake. Intrigued by this phenomenon, German researchers at the Max Planck Institute of Animal Behavior in Radolfzell/Konstanz and the Cluster of Excellence Centre for the Advanced Study of Collective Behaviour at the University of Konstanz have carried out experiments to determine whether cows, sheep, and dogs can detect early signs of earthquakes. The researchers attached sensors to animals in an earthquake-prone area in northern Italy. Their movements, which were monitored for several months, indicated the animals did, in fact, become unusually restless in the hours before an earthquake happened. What's more, the closer the animals were to the epicentre of the impending quake, the earlier they grew restless.[1] Research to learn more about the how and why of this phenomenon continues.

If any of the miners in the No. 2 colliery this evening noticed there were fewer rats than usual scurrying around, no one commented on it. That's hardly surprising, for an absence of rats in the mine wasn't something the men would have noticed immediately.

With supper eaten, the men in the No. 2 mine returned to work. The sweating, swearing, and singing resumed. The coal pans continued to clatter. The rakes creaked and groaned as they hauled up to the surface the tens of thousands of tonnes of coal that came out of the No. 2 each shift. It was situation normal, just another busy evening in the mine. At least, it was until shortly after the supper break ended, a few minutes before seven o'clock. That was when the earth groaned and moved ever so slightly, like a sick old man rolling around in his bed. According to Louis Frost, DOSCO's chief mining engineer, "[This bump] was felt generally over the district. But on examination there was no physical evidence to show where it had occurred or what damage, if any, it had caused."[2]

This tremor, which seemed like any of the other ones that plagued the mine—more than five hundred of them in recent years—brought large chunks of coal and rock cascading down from the coal faces on all three levels of the mine. It also clouded the air with blots of coal dust. But that was nothing unusual, nothing to be concerned about. "We had those little bumps all the time. I never paid them much attention, and I didn't even feel this one," said Harold Brine.[3] The casualness of his dismissal of the threat the bumps posed is striking. It was also typical. Some miners welcomed these small bumps, for they loosened windfalls from the coal face and made it easier to take down what remained lodged in place. That was always a good thing; the greater the tonnage of coal the miners sent to the surface each shift, the more they earned.

Bowman Maddison, who was labouring on the pack timbers being constructed on the coal face at the 13,400-foot level, wasn't among the men who welcomed a bump or who ignored it, no matter how slight the tremor. When he felt this one, he stopped cold.

"Did you feel that?" he asked his co-worker.

Caleb Rushton nodded. It was with hearts pounding and sweaty palms that the two men held their breath and waited anxiously several moments for whatever would happen next. They felt as helpless as they were. When nothing happened, both men exhaled. Then they laughed nervously. Buoyed by a vague, tenuous feeling of relief, they went back to work. Like many of his fellow miners, Bowman Maddison had a fatalistic attitude to working in the mine. "When my time comes, I imagine I'll get it," he often said.[4] This time around, he decided, he'd dodged yet another bullet.

Whenever a strong bump rattled the mine, safety protocols dictated that all underground work stopped for twenty-four hours. No work meant no pay for the miners. That was something none of them ever wanted. For Maddison, Rushton, and all the other labourers, a lost shift meant $9.74 less on their weekly paycheques.

That was a substantial financial hit, and so, like their co-workers, the men were relieved anytime work wasn't suspended after a bump.

Elsewhere on the 13,000-foot level of the mine at seven o'clock, a couple of technicians from the provincial Department of Mines were tinkering with monitors they used to measure the stresses the shifting of the earth exerted on the mine. Just as some of the miners had, the researchers froze. The instant they felt the tremor, the hair on the backs of their necks bristled. So, too did that of shift overman Charlie Burton, who paused to jot down in his notebook the time the earth shifted. A veteran miner, Burton had survived the 1956 explosion in the No. 4 mine and had emerged as one of the heroes of the rescue mission that saved the lives of eighty-eight men. By 1958, he was operating a compressed-air chipper pick and serving as an overman in the No. 2 mine. As a supervisor, he had to record the details of any other relevant events that occurred, especially any that affected production or workplace safety.

Miner Cecil Colwell cursed his bad luck when he had to leave work because of a cut on his arm. When he found the first aid station closed, he returned home to have his wife help him clean and dress his wound. Colwell had no way of knowing it, but although his bloody misfortune cost him four hours' pay, it saved him from a world of pain and suffering. It may well have even saved his life.[5]

CHAPTER 9

NOTHING TO WORRY ABOUT

Thursday, October 23, 1958
8:00 P.M.

An hour after the mini-bump rattled the No. 2 colliery, it was as though nothing unusual had happened that evening. The mine's powerful ventilation fans had cleared the air of double trouble—firedamp and coal dust.[1] Both are bad news in any coal mine. Both can kill.

Firedamp, which is colourless, odourless, and highly flammable, is a by-product of the process that causes plant matter to decompose. As such, pockets of the gas accumulate naturally in coal seams and shale deposits. They are released into the air whenever miners encounter them, and that's when problems occur. Firedamp is largely made up of methane (CH_4). If you studied chemistry in high school, you'll know methane is a hydrocarbon that's an integral component in the natural gas many of us burn to heat our homes and power our stoves and water heaters. Methane is prominent on the list of notorious greenhouse gases—along with carbon dioxide, nitrous oxide, and fluorinated gases—that are driving climate change.

Because firedamp is less dense than air, it accumulates near the roof whenever there's inadequate ventilation in a coal mine. That's problematic for anyone who's walking upright. Inhaling firedamp-laden air makes you woozy or can even knock you out. If the concentration of the gas is intense, prolonged exposure can be fatal. No less worrisome is the other danger firedamp poses to coal miners: when the concentration of the gas in the air reaches 9.5 percent, a flame or even a spark from a machine or a metal tool as it strikes stone can cause an explosion. The people of Springhill knew about that through bitter experience. Firedamp had fuelled the catastrophic 1891 explosion that rocked the town and killed 125 miners.

Coal dust, that other killer of coal miners, also causes explosions. In a confined space, any combustible dust that's in high concentration can ignite with deadly results. Without adequate ventilation, dust explosions can be a major hazard in coal mines, grain elevators, and other industrial environments. Given their fiery combustibility and the fact that dust explosions can be safely contained in controlled conditions, they're a favourite tool of special effects artists and filmmakers. Weapons makers also like them. Thermobaric weapons—which are also called aerosol bombs—rapidly saturate an area with an easily combustible material that's then ignited to produce an explosion of devastating force. Such weapons are the most powerful non-nuclear weapons in existence.

However, on the evening of October 23, 1958, the ventilation fans in the No. 2 mine were doing their job in the wake of that seven o'clock mini-bump. With the dangers posed by coal dust and firedamp seemingly under control, work had resumed in earnest. It was as though nothing out of the ordinary had happened that evening. That was Maurice Ruddick's attitude.

As always, that grim awareness was shelved back-of-mind for all the men who were working the afternoon shift in the No. 2 mine. Maurice Ruddick was the "top man" in the crew chipping away at the top end of the coal face at the 13,000-foot level. The shiny chunks of coal he dislodged tumbled into the coal pan below. Ruddick was good at his job, and he enjoyed it. He sang a happy tune as he swung his pickaxe and sweated.

A short distance away from Ruddick and below him, Harold Brine, the "second" miner in this same segment of the coal face, was also wielding his pickaxe and shovel. Working like cogs in a well-oiled machine, they could keep the hopper of the coal pan full. Brine, Ruddick, and the other miners above and below them on the coal face were hard at it on the 400 feet of the wall that ran between the 13,000- and 13,400-foot levels of the mine. As they worked, the noise could be deafening at times. At other times, the mine was quiet as a tomb. That quiet began when oversized clumps of coal jammed the pans, or if the engine that rattled the pans and kept the flow of coal moving broke down. When that happened, the pans all along the level stopped, and work slowed or halted. Until the problem was fixed, any coal that fell from the coal face and onto the stilled conveyor would inevitably spill onto the floor. There it would pile up, or else the weight of the unbalanced load sometimes caused the conveyor pan to list dangerously or even to topple over. Regardless, any stoppage was a problem that had to be fixed as quickly as possible.

All the miners working the No. 2 mine counted on smooth, continuous operation of the pans in order to do their jobs and earn their money. Doing so was difficult enough at the best of times. Rivulets of soot-blackened moisture streaked down the men's shirtless torsos; perspiration beaded their brows. In the bobbing light cast by the miners' lamps, the whites of their eyes glistened like stars in a cloudless night sky.

When Harold Brine looked up to assess the progress he and Maurice Ruddick had made on the coal face this evening, he smiled. However, Ruddick was now carving his way through a vein of fine, crumbly coal, and this made for slow going. Brine knew he had a few minutes to take a breather. While another man might have sat down, Brine was no slacker. He wasn't inclined to be idle or to sit around chatting if there was work to do. Brine and the other men in his crew worked cooperatively. They were paid according to the volume of coal that went to the surface "on their tally." Time was money.

Tonight, Brine began helping the labourers who were moving around lengths of wood that would be used to build the props holding up the roof. The labourers—Garnet Clarke, Currie Smith, and Herb Pepperdine—were all good guys. Harold Brine liked and got along well with them. He was happy to give them a hand piling the timbers near where they'd be needed. Why not? It was the job of the miners themselves to position the props in place just behind the area on the coal face where they were working; the miners knew best where support was needed to keep the roof from collapsing on them.

Once the labourers had moved all the available timber to where it needed to go, their work came to a temporary halt. The men stood around while they waited for more wood to arrive. Of course, as inevitably happened whenever there was a few minutes to spare, some of the guys began talking and joking around. Garnet Clarke was just starting to tell his buddies a dirty joke when it occurred to Harold Brine that rather than stand there listening to a bad stand-up routine, he could make better use of his time if he went to see what other work needed to be done below, at the top end of the coal face on the 13,400-foot level. With the pans stopped down below, it was unusually quiet. Time was money to a miner, and so Brine was eager to discover what the problem was; he wondered if there was something he could do to help get things moving again.

"Maurice, I'm headed down below," he called out. "I'm going to help clean up some of the spillage."

"Okay. You go ahead," said Ruddick. "I'll finish what I'm doing here and then put in a prop or two."[2]

With that settled, Brine picked up his shovel and handpick and set off for the 13,400-foot level. The walkway down the head to the level below was uneven and ran at about a 20-degree slope. A man had to be mindful of where he stepped. Along the way, Brine greeted and exchanged pleasantries with those men he knew who were working this section of the coal face.

With every step he took, Brine was descending ever deeper into the bowels of No. 2 mine as the minutes were ticking away before all hell would break loose.

THE ENGINE THAT rattled the pans and kept the flow of coal moving along the coal face at the top end of the 13,400-foot level still wasn't operational when Harold Brine arrived. Despite this, the two dozen or so men working there were busy. Brine set to work helping the labourers who were cleaning up coal that had overflowed the idle pans and was mounding up on the floor of the mine. That done, he climbed up the coal face and reflexively began chipping away at some loose coal. Force of habit. When he realized what he was doing, he paused to look around. He saw that Gorley Kempt, the operator of the electric motor that powered the coal pans machinery, was talking with labourers Caleb Rushton and Bowman Maddison. Those two men, both company hands, had teamed together for about a year "on the timber load." Brine had worked with all three men for several years and knew them well, although none of them was a close friend.

Kempt, a thirty-eight-year-old native of the nearby mining town of Joggins, hailed from a mining family. He'd worked as a coal miner for most of his adult life. In 1956, he'd been among the barefaced miners who'd risked their lives to venture down into the No. 4 mine to rescue survivors of the explosion that had claimed the lives of thirty-nine men. Noxious gases had rendered

Kempt unconscious four times, but each time he'd regained consciousness, he'd rejoined the rescue mission. Kempt was fearless, or maybe it simply wasn't in his personality to quit. He was a lean, chain-smoking extrovert, a bundle of energy who laughed easily and often.

A carpenter by trade, Kempt had been just twenty-one and was unemployed when he wed fair-haired Springhill girl Marguerite ("Margie") Cliffe. That was in June 1941. Six months later, the couple welcomed a son into the world, the first of their two children. "I was too young to marry [Gorley] at sixteen," Margie Kempt would muse many years later. "But everybody was doing it—marrying a miner. Well, we didn't have much choice but to marry a miner. Young, strong, fine-looking fellows they were, and [they had] the best pay package in town."[3]

Many people in Springhill would have told you that Gorley Kempt was one of those "fine-looking fellows." He was a gad-about who liked nothing better than to race around Springhill on his motorcycle. Like most miners, Gorley enjoyed an occasional drink (or two), and he loved to play cards and go fishing. During the annual summer break at the mine, he and Margie entertained other couples at the Kempts' summer cottage, a jury-built structure on the Minas Basin, an inlet off the Bay of Fundy. Ever gregarious, Gorley was a charmer, and he spread merriment wherever he went. He was everybody's friend, whether they were male or female.

The two men Kempt was speaking with when Harold Brine arrived at the 13,400-foot level were best pals. Caleb Rushton and Bowman Maddison were also birds of a feather. Rushton, at age thirty-five, was quiet and studious, and although he was divorced and had remarried, he was a devout churchgoer. He was something of a rarity among miners for two reasons: he was a high school graduate, and he loved to read. He spent much of his spare time poring over books he borrowed from Dr. Burden. Rushton had worked three years in the mine's lamp cabin; then in 1957, despite the fact he hated working underground, he began doing so because he needed the money.

Rushton's pal, Bowman Maddison, was a forty-two-year-old father of three children. Like Rushton, he was studious and soft-spoken. In large measure, that was because his parents hadn't been married when he was born. That stigma had scarred him for life, and he wasn't hesitant to admit as much. "It bothered me [and] kept me from going and mixing with people," Maddison said.[4] In addition to being an avid reader, he was a philatelist, a stamp collector. Maddison, who liked to tell people, "My people in the old country were musical,"[5] also wrote pop songs. He sometimes entered them into competitions; other times, he sent them to well-known singers whose vocal stylings he admired. In fact, just the day before he had mailed off a tune to Juliette ("Our pet, Juliette") Cavazzi, the singer who hosted a popular musical variety show on CBC Television that aired on Saturday nights, right after the hockey game. Maddison fancied that his latest tune, a wistful love song titled "Dearest, My Heart Is Calling," might be one that Juliette would perform on the show. Why not? Maddison figured it never hurt to dream about one day penning a hit tune that hope-fully would earn him enough money to escape minework.

Gorley Kempt, Caleb Rushton, and Bowman Maddison might have been talking about their favourite pop tunes, or they might have been trying to figure out what mechanical issue had stopped the pans. However, it was just as likely that Kempt was telling a joke or spinning a yarn; he always had a ready supply of both on the tip of his tongue. Whatever they were talking about, Harold Brine saw that the three men were deep in conversation; Caleb Rushton had settled down and was sitting with his back against the coal face, his legs stretched out before him. Everything seemed as it should be. But looks can be deceiving.

CHAPTER 10

THE BIG ONE

THURSDAY, OCTOBER 23, 1958
8:06 P.M.

THE RESIDENT MANAGER OF DOSCO's No. 2 MINE WAS AT home when the earth moved. George Calder and his family lived in one of the mansions the company provided to the on-site mine executives; the dwelling, one of the largest in town, stood barely 50 yards from the archway at the head of the roadway that led to the mine. The Calders had just finished supper and were clearing away the dishes when the house began to shake.

At age forty-eight, Calder was an experienced mining man, and so he knew instantly there was big trouble at the mine. A native of nearby Joggins, he was one of ten children—seven sons among them—of a coal mining family. His father had worked underground until he was seventy-five, long after the time when most men called it quits or were dead as a result of black lung or one of the many other ailments that killed so many coal miners. Scottish-born John Calder had immigrated to Canada in his early teens and immediately found work in the Joggins mines. Conditions there were as hazardous for the miners as they were difficult; the coal seams were narrow, and they dropped off at an acute angle of almost 45 degrees. The men worked while crawling on their stomachs in a space too small for even a child to stand. Joggins miners dislodged coal using a small pick, and they then collected it with their bare hands. It was a brutal way to make a living—too much so. John Calder moved his growing family to Springhill, where work in the mines was "easy." Relatively speaking, at least.

John Calder was highly intelligent, bookish, and self-educated. He also knew the value of going to school; however, he and his wife had no money to pay for their children to attend university or technical college. That stark reality was made clear to George Calder the day in 1929 that his mother took him aside to explain, "There's no money for you to go on in school. Your dad is going to take you over to the mine tomorrow to introduce you to some of the managers. You're going to start working in the mine."[1] And that's what he did. At age nineteen, George Calder became a miner, and each Friday he turned over his paycheque to his mother. The money helped feed the family

Like his dad, George was a big man, physically robust, God-fearing, and whip smart. Not surprisingly, he rose steadily through the ranks of management at the Springhill mine. He served as an underground shift boss for eight years, then after a brief stint as assistant manager of the No. 2 mine, he became manager—"the Number One guy." It was a job for which Calder was well suited, for as one observer noted, he was "widely admired and respected by all the workmen for his fairness, intelligence, and courage."[2]

Calder oversaw day-to-day operations in the mine and made sure everything was running efficiently and as safely as possible. He'd gone down into the mine earlier that day for a look around and found that everything seemed normal. "The day was going well, and the mine was in good order," he would later state.[3] Despite this, when at 8:06 p.m. Calder felt the second tremors of the evening, he instinctively knew this was no ordinary bump. Snatching up the phone, he repeatedly called the mine foreman who was on duty that evening. When no one answered, Calder knew his instincts were right. He was still donning his hat and coat as he rushed out the door. Calder was headed to his office, which was in the big wood-frame building right next door to his home.

Springhill mine manager George Calder in his later years. (Courtesy of Peter Calder)

ALMOST AT THE same moment that George Calder felt the earth move, over at the Springhill Medical Centre on Main Street, a block east of the DOSCO offices, Dr. Arnold Burden was finishing up with a patient. Like George Calder and most everyone else in town, the doctor knew a big bump when he felt one. In Burden's mind, he was momentarily transported back in time to the battlefields of WWII. "I felt three distinct shockwaves, like a stick of bombs being dropped from a fighter bomber, exploding just fractions of a second apart," Burden would say. "We all knew from the magnitude of the impact that it was a major disaster at the mine."[4]

Beverly Reynolds sensed the same thing. Reynolds, who was the sister of the medical centre's nurse and the wife of a miner, was working in a back room of the downstairs drugstore when the Bump hit. The tremors were so strong, they knocked her off her chair, and she tumbled onto the floor. She had heard and felt bumps before, but never one with this intensity. After picking herself up, she rushed upstairs to search for her sister. "Oh, my God! Wes is down there," she shrieked before turning and running back down the stairs.[5] Reynolds was headed to the pithead in hopes of finding out if her miner husband was all right. The knot in her stomach told her something very bad had happened.

Arnold Burden and his colleagues, Drs. DeWenten Fisher and Carson Murray, shared Reynolds's fears. They all knew if this bump was as bad as they feared, a lot of injured miners would soon be arriving at the town hospital. With that in mind, the doctors hastily shooed patients out of the waiting room and closed the medical centre. Burden raced to the mine while his three colleagues made their way to All Saints' Cottage Hospital, where emergency preparations to receive injured miners were already under way.

The moment hospital administrator Stanley Tibbetts had felt the Bump, he knew there was big trouble at the mine. His worst fears were confirmed when he called the DOSCO offices and no one answered. That confirmed in Tibbetts's mind that he had no time to waste; he'd had lots of experience dealing with crises at the mine. That being so, he and nursing supervisor Rebecca Hargreaves held a quick meeting at which they decided to immediately discharge any patients who were medically stable and could safely be sent home. Tibbetts and Hargreaves wanted to free up as many of the hospital's fifty-six beds as quickly as possible. Even more sobering was the decision the pair made to ask the head of the local militia unit to open the barn-like tin-roofed building on Drummond Street that served as the town armouries for possible use as a temporary hospital and as a morgue, if one was needed.

While all this was happening at All Saints' Cottage Hospital, over at the town hall, Mayor Ralph Gilroy had been chairing a town council session when the meeting room started to shake. The mayor and five councilmen, all of whom were usually as voluble as crows, fell into stunned silence; they gripped their seats and hung on for dear life. Once the tremor ceased, they hustled out of the building.

At the mayor's house a few blocks away, his wife bolted out the front door. Like many of her neighbours, Grace Gilroy feared an earthquake was about to bring the house down upon her. Better to be out in the open air than huddled inside, she reasoned.

As Grace Gilroy was rushing into the street, a few blocks to the east, thirteen-year-old Anne Murray was bidding a girlfriend goodnight; the two girls had just watched a television program together. Murray, the Springhill-born daughter of one of the town's four doctors, was destined to enjoy a phenomenally successful career as a pop singer and to be lauded as Springhill's most famous native; however, in 1958 she was just one of the girls in grade nine at Springhill High School. Murray had just closed the front door when the Bump happened. "The rumbling was so loud mum called out from upstairs to ask if I was all right," Murray recalled. "She worried that I'd fallen down the stairs."[6]

At 8:06 p.m., over at the Springhill Armouries, seventeen-year-old Billy Kempt and his militia mates were waiting to unload supplies from army trucks from nearby Amherst. The trucks were due to arrive at any minute. The young soldiers had just assembled and were on parade inside the armouries when the earth trembled beneath their feet. "I saw the building shake, and my first thought was that one of the army trucks in the convoy had missed the turn and crashed into the front of the armouries," Kempt remembered.[7] In the next breath, he realized what had really happened.

Springhill born and the son of coal miner Gorley Kempt, Billy knew all about the seismic quakes that periodically rattled the town and its mine. "They were always in the back of your mind, even when you weren't thinking about them," said Kempt.[8] And so, he understood that this was no ordinary bump. So did Norma Ruddick and her neighbours at the lower end of Herrett Road, on the land to the west of town.

That thoroughfare, well travelled yet still rutted and weedy, was literally a stone's throw west of the pithead of the No. 2 mine. Many of the families in Springhill's small, vibrant, and closely knit Black community lived along lower Herrett Road. These "rowers"—as other townspeople called them—endured the discomforts of the drafty, poorly insulated dwellings the mine company had erected there in the early years of the twentieth century.

The evening of October 23 had been unseasonably warm, and as darkness fell, several of the older Ruddick children had stayed outside, playing with their friends. "But suddenly the earth began to shake," remembered Valerie (Ruddick) MacDonald, who was seven at the time. "We knew right away there was trouble at the mine, and Dad was at work.

"When we ran into the house, Mum was in the living room, sitting in her favourite chair with one of the babies on her lap. She had the saddest look on her face. The Don Messer show had been on television, and the power went out. When it came back on a few minutes later, Marg [Osburne] and Charlie [Chamberlain] were singing a hymn: 'Farther along, we'll know all about it / Farther along, we'll understand why / Cheer up, don't worry, live in the sunshine / We'll understand it all by and by. . . .' That song gave Mom some comfort, but she was in a tearful state. She knew something bad had happened at the mine."[9]

News of the Springhill mine disaster made the headlines in the *Halifax Chronicle-Herald* and in newspapers across Canada. (Ken Cuthbertson)

CHAPTER II

AN UPHEAVAL, NOT A CAVE-IN

Thursday, October 23, 1958
8:06 p.m.

The coal pans still weren't working, but Harold Brine liked to keep busy. He was chipping away at some loose coal at the top of the 13,400-foot coal face when a co-worker interrupted his labours. The light on Eldred "Ed" Lowther's cap flashed brightly in Brine's eyes as Lowther rested a foot on the side of the coal pan and leaned in to speak into Brine's ear. Forty-five-year-old Lowther, who'd worked as a coal miner for twenty-eight years, knew everything there was to know about working underground. When he spoke, other miners usually listened.

"Did you just feel a bump coming down the wall?" Lowther asked Harold Brine. His voice was edged with concern. Or perhaps it was fear.

"No," replied Brine.

"Well, it shook everything."[1]

Brine was about to reply, but at that exact moment "she let go," as the miners who survived their ordeal would later say.

The Big One—the killer bump everyone in Springhill had

always feared and knew in all likelihood would happen one day, raced along the working seams of coal. It radiated upward from the depths to the pithead and then out across the landscape and into the town like a runaway locomotive.

The layperson's perception of a mine disaster is that the roof or walls give way, and miners die in a cave-in. Those who are lucky to survive that initial trauma find themselves in a nightmarish scenario: most likely they're injured, alone in the eternal darkness, struggling to breathe, and entombed under tons of fallen rocks and soil. However, owing to the peculiarities of Cumberland County's geology, there was no cave-in at Springhill's No. 2 mine. That wasn't what killed men. It was an upheaval, not a cave-in, that killed them.

The area between the roof of the mine and the surface far above was mostly filled with sandstone. That sedimentary rock formed 400 million years earlier, around the same time as the coal seams that marbled it. Such a geological formation is tremendously strong; it can span a sizable gap without giving way or collapsing. Before any mining was done in the No. 2 mine, the cumulative weight of thousands of feet of sandstone and soil had rested securely atop and around the coal seams. It had done so for eons. When miners removed coal from the coal faces being worked in the autumn of 1958, they carved out pockets of empty space. The roof of the mine in these areas was supported by the rows of packs the miners erected up and down the three coal faces.

When those coal faces were staggered step-like, the geological stresses created by mining activities were problematic. But all things considered, the small bumps that happened with such dismaying regularity generally didn't pose any major or overriding safety concerns. That risk changed as the mine went ever deeper. The dynamics of the geology below 10,000 feet were different. As the stresses grew, so too did the threat of major seismic disturbances. That was especially so when the miners brought the coal faces at 13,000, 13,400, and 13,800 feet into an alignment that was more or less vertical. While the mine floor along the coal faces continued to slope at about a 20-degree angle, the run between levels was now an awe-inspiring 1,200 feet—almost four Canadian football fields long.

DOSCO's managers for many years had been intent on aligning the coal faces in the No. 2 mine. However, as mentioned earlier, the impetus to do so and to increase coal production had taken on fresh urgency following A.V. Roe's 1957 purchase of DOSCO. In retrospect, it seems apparent this was akin to running faster and faster on the spot. However, none of that much concerned DOSCO's chief mining engineer, the man who brought about alignment of the coal faces in the No. 2 mine. Louis Frost knew only that his boss, DOSCO general manager Harold Gordon (no relation to A.V. Roe president Crawford Gordon), was demanding to see an increase in coal production in Springhill.

Frost, sixty-one years old and Scottish born, was a chain-smoking bundle of energy. He'd started his mining career in his native land when he was fifteen, served in the British army during WWI, and then earned a degree in mining engineering from Heriot-Watt College in Edinburgh. Frost had immigrated to Canada in 1927. After joining DOSCO as a ventilation and field engineer, he spent three years as the assistant chief mining engineer and took on the top job in 1930. He was only thirty-three at the time, but unlike many of the senior people in Nova Scotia's coal mining industry, he had academic credentials; that distinction won him instant respect and credibility.

By the mid-1950s, Louis Frost had earned a reputation as one of Canada's most knowledgeable mining experts, and he was overseeing operations at DOSCO's coal mines in Springhill, Pictou County, and Cape Breton. Diminutive and by that time balding, he bore a facial resemblance to Yul Brynner, the Hollywood film star of yesteryear. And like Brynner's regal character in the classic 1956 film musical *The King and I*, Frost didn't lack for self-confidence or assertiveness.

After studying the available data and visiting the Springhill mine numerous times, Frost had concluded the best way to increase the No. 2 mine's productivity and to reduce bumping was to double down on the strategy of bringing the three coal faces into alignment. "We had a bump on October 9, 1957, and we attributed [it] to the walls being out of line," he would later explain. "We held discussions with everyone concerned, and we agreed that the proper thing to do was to try to bring the walls into alignment as quickly as possible."[2]

Work to bring that about had picked up speed late that autumn, and while there were no bumps in the first two months of 1958, a sizable jolt had rocked the mine in March, killing one man. Then there were thirteen bumps in the four months from April through July. Not surprisingly, a growing number of miners began to question the wisdom of Frost's strategy. When miners' union officials urgently requested a meeting with management, one was held at the DOSCO offices in Springhill on September 20. Frost hadn't attended. He'd left it to mine manager George Calder to sit down with union officials, listen to their concerns, and then explain to them that work to align the three coal faces—on the 13,000-, 13,400-, and 13,800-foot levels—would continue apace. They were now within 15 feet apart.

"Regardless of who suggests a method of work, whether the government, the workmen, or any other agency, we will not change our method of work without a very complete and thorough investigation, and when we are convinced that the proper step to take is to change that method of work, we will certainly do so," Frost explained.[3]

Union officials and the rank and file—among them Harold Brine, Maurice Ruddick, Tommie Tabor, and other veteran miners—weren't convinced that "the college boys" had it right. As the wife of one ill-fated miner would note, "The men didn't want to go to work. [But] if they didn't . . . they didn't get any pay, so they were more or less forced into it."[4]

WHETHER THE MINERS' fears about aligning the coal faces were valid remains a topic for debate. What doesn't is that the bump that happened on the evening of October 23, 1958, was a disaster of devastating proportions. When the subterranean thunder—the product of three distinct shock waves—rumbled through the earth beneath Springhill, many men died instantly. They never had a chance to escape. What triggered these killer quakes is impossible to say, other than to speculate they were most likely the result of natural forces that were as unpredictable and mysterious as those that trigger any earthquake. Or it could be that what touched off what to locals became known as "the Bump" was the removal of a large mass of coal from one of the seams or coal faces some-where in the mine. Whatever the reason, once the earth began to move, there was no stopping it. "It happened so fast that I couldn't tell you what happened," a lucky miner who made it out alive would say.[5]

The seismic forces unleashed in the earth's depths were so sudden and powerful that—counterintuitive though it sounds—the floor of the No. 2 mine at Springhill heaved *upward*. When it did so, it dashed miners and everything around them in the mine against the roof, crumpling coal pans, trolley cars, steel rails, stone-and-wood support packs, and all other mine infrastructure. At first glance, the Bump's impact seemed to have been puzzlingly, maddeningly, and brutally inconsistent. But it wasn't. As the ground heaved, the coal faces in the No. 2 mine shot upward and outward, like a giant desk drawer that suddenly springs open with killing effect.

Many of the miners who survived the Bump did so only because they chanced to be standing in the right spots when the earth moved. Instead of being crushed instantly as their co-workers on the coal face were, the men working at the top and bottom of a 400-foot run of the wall, closest to the levels above and below, had the best chance of survival. The seismic energy that surged up from below sent many of them flying, and the mighty gust of compressed air that then rushed in to fill the vacuum created by the displacement of the coal seam hit them. It was where a man was standing or working when this one-two punch hit that dictated whether he lived or died.

The vast groundswell of unleashed energy resonated along the coal faces. It was most intense in the deepest levels of the mine. It was so strong at the 13,800-foot level that "words can't describe it," as one of the thirty-four miners who were working there would relate. "There was a terrific jolt that [stunned] everyone. For a few minutes we didn't know where we were."[6] Harold Cummings, another of the few men to survive at this level of the mine, would tell a newspaper reporter, "That [was] the first bump I didn't feel or hear in thirty-eight years [on the job]. I was working about one hundred and fifty feet from the face of the 13,800 coal face. When I came to, I was . . . in hospital."[7]

Tommie Tabor wasn't as fortunate. Like many of his workmates on the coal face at the 13,800-foot level, he died instantly. The second the mine floor heaved upward, it crushed everyone and everything against the roof with no warning. The cause of Tabor's death as recorded on the death certificate that Dr. DeWenten Fisher would fill out was "severe multiple contusions and crushing."[8]

All those men in the No. 2 mine who escaped death from the Bump faced a second grave threat—the danger posed by firedamp, the miasma that has always been the silent killer in coal mines everywhere. When all was well in the mine, a huge ventilation fan at the pithead blew fresh air down into the depths and sucked out the noxious fumes. Down below, as a matter of course, workers routinely opened and closed wooden trapdoors that directed fresh air to all levels of the mine. The Bump wrecked some of those doors and closed off sections of the passageways where fresh air would normally have circulated. The result was a buildup of fire-damp that posed a lethal threat to the survivors of the Bump and to the fearless barefaced rescuers and the draegermen—specially trained miners who donned scuba-like air tanks and other gear—who went into the mine looking for them.[9]

It would be long, agonizing days, weeks even, before the families of many of the missing men received official word that the remains of their loved ones had been located. Sometimes the wait to recover a corpse and return it to loved ones for a funeral took even longer. The work of positively identifying bodies was a difficult and gruesome task. As often as not, final identification of those men who'd been crushed was almost impossible, and in some cases was accomplished only with the help of dental records and other identifying bodily features—and the unretrieved brass tags in the lamp cabin.

"I NEVER KNEW a bump could be so hard," said Fred Hahnen, one of the few survivors among the crew of thirty-four men who were working the coal face at 13,800 feet when the mine floor heaved.[10]

Four hundred feet above, Harold Brine and co-worker Ed Lowther were lucky to escape immediate death. However, the jolt knocked both of them senseless. The force of the Bump hurled their limp bodies through the darkness, bouncing them off the mine roof before body-slamming them to the mine floor. Miraculously, both men survived with only minor injuries. And so did Gorley Kempt, Caleb Rushton, Bowman Maddison, and seven other men who were working on the 13,400 level, at the bottom end of the 13,000-foot coal face—Hughie Guthro, Joe Holloway, Wilfred ("Fred") Hunter, Larry Leadbetter, Joe McDonald, Ted Michniak, and Levi Milley. Collectively, these men would become known to would-be rescuers and the media as "the Twelve."

THE DEATH, DESTRUCTION, and devastation wrought by the Bump shattered the No. 2 mine. To be sure, none of the men working there on this night totally escaped its impact. Many of them suffered physical injuries. Others were scarred mentally. No one who survived would ever forget the horrors of this day.

Maurice Ruddick had continued working at the top of the 13,000-foot coal face after Harold Brine's departure. At the moment the earth moved, Ruddick was finishing the work he'd been doing and was about to join Brine at the bottom end of the 13,000-foot wall. He saw and heard nothing when the mine floor jumped up at him. The force of the upheaval sent him flying through the dark like a raindrop in a gale.

Garnet Clarke and Herb Pepperdine, the labourers whom Harold Brine had helped move timbers around only a few minutes earlier, were working on the coal face a couple of arm's lengths below Maurice Ruddick. The upheaval hurled Clarke backward, landing him on top of Pepperdine. The two men went down in a crumpled heap. Currie Smith, who'd been working beside them, fared better only because of the luck of the proverbial draw.

"I was just sittin' there waitin', and all at once the pans started shakin'... umph! She went off," Smith would say. "[The Bump] just kinda knocked my cap off. I grabbed the [lamp] cord and pulled it back on."[11] Although the tremors had staggered him, Smith never lost his footing. And while he instinctively knew there'd been a bump, in the pitch black, he initially had no idea how big or bad it had been. Nor did he know how hard it had hit his co-workers, Frank Hunter (Fred's twin brother) among them.

Hunter suffered ruptured eardrums. How and why, he would never know. About all that he could say for sure and all that mattered at the time was that the pain was beyond excruciating. It felt as though someone had jabbed the sharp end of a pencil through both his eardrums, and blood began trickling from his ears. What caused this injury was the abrupt change in air pressure that swept through the mine when the earth moved. "It just knocked you out, the concussion. . . . Just like someone [had] fired a shotgun near your ear," Herb Pepperdine would recall.[12] Why some men suffered such devastating impacts while others escaped serious injury is impossible to say for sure.

Doug Jewkes had been standing next to Frank Hunter, between him and another miner, when the Bump hit. All three men were working on the coal face. The force of the upheaval knocked Jewkes six feet through the air like a rag doll and left him half-buried in coal. Unlike Hunter, his eardrums were unaffected. As for the man who was standing next to Jewkes, we'll never know the specifics of his injuries, only that he died instantly under a deluge of coal that was more than heavy enough to break your limbs or fracture your skull. More than half the miners who were working on the 13,000-foot level of the mine at 8:06 p.m. shared that same crushing fate. The men's lives ended as quickly as if they'd been struck by lightning. Then, there was the luckless Percy Rector, the man operating the motor that powered the pans along this stretch of the coal face. His fate was nightmarish, infinitely worse than immediate death.

Rector was that fifty-five-year-old miner who'd forgotten his identity tag at home on this day—the only time in the thirty-seven years of working in the mine that he'd ever made that mistake. As a result, he'd been obliged to go to the mine manager's office to ask for a requisition slip that gave him permission to sign out the cap light he needed if he wanted to work the afternoon shift. Without the light, Rector could not have gone into the mine. It would have been much better for him if he'd gone home instead.

The instant of the Bump, Rector had the misfortune to be working near a roof pack that had been under construction. The upheaval of the floor flung him backward, sending him crashing into the half-built pile of stone and timber at the exact moment the mine floor was shooting upward. As the pack crumpled like a house of cards, Rector's right arm was caught a couple of inches below the elbow, sandwiched between two thick chunks of wood just as millions of tons of earth, coal, and sandstone came crashing down.

To analogize, it must have felt like the pain of having a truck door that can't be reopened slam shut on your fingers. Only ten times worse and with no end to the suffering. In the pitch black and with no way to lift the colossal weight off his arm or to saw through the heavy wood that gripped him, Percy Rector was pinned, hideously and hopelessly; without a blood supply, a human limb becomes unsalvageable after about six hours.

There was only one way his workmates could free him, and that was to hack off his arm using whatever tools they could find. And they'd have to do that without the aid of any anesthetic. It was a textbook example of the infamous Hobson's choice: If they did nothing, Rector would surely die, later rather than sooner. And since none of Rector's workmates had medical training and they had no surgical equipment at hand, they knew if they tried to cut off his arm, the trauma and blood loss surely would kill him. The decision the men faced was a life-and-death one for which there was no easy—and no real—resolution. Unless rescuers appeared soon, the men would need to decide what to do. That was a decision none of them wanted to make.

CHAPTER 12

CONFUSION REIGNS

Thursday, October 23, 1958
8:15 P.M.

By the time Dr. Arnold Burden left the medical centre, the streets of Springhill were full of people and cars. Main Street was as busy as it was on a typical Saturday, the day when most townspeople did their weekly shopping. However, on this Thursday evening, in a town in which extended families were large and the community was so tightly knit, as the *Springhill-Parrsboro Record* would report, "Within minutes of the terrific bump in No. 2 mine that had rocked the town and surrounding countryside . . . the scene at the pithead was a duplication of that which followed the No. 4 mine explosion on November 1, 1956. Anxious to discover the fate of their loved ones or friends, people . . . had hurried to the surface plant, there to await the news from the depths of the deepest coal mine in North America."[1]

Zora Maddison, the fourteen-year-old daughter of miner Bowman Maddison, joined that frantic rush to the mine. She had been at home with her mother when the earth shook; both mother and daughter knew instantly what was happening. Zora didn't hesitate, not for a moment. She was pulling on her jacket even as she rushed out the front door and set off running to the mine as fast as her feet would carry her. "I'm going to the mine!" she called back over her shoulder.

A few blocks away, when the Bump caused her house to shake and groan, Margaret Guthro was in her kitchen, setting up to give a girlfriend a home permanent. She immediately dropped a fistful of curlers, for the tremors brought a gut-wrenching feeling of déjà vu. She still had vivid memories of the three uncertain, agonizing days when her husband, Hughie Guthro, had been trapped underground in the No. 4 mine back in 1956. "Oh, my God! This can't be happening again," she cried. But it was, and so Margaret Guthro was frantic as she too lit out for the mine. By the time she got to the pithead, a large crowd of men, women, and children had already gathered outside the entrance to the hoist shed. They had come running from all directions, and more people were arriving with every passing minute. Great swirling clouds of black coal dust, like the great fogs of Dickensian London, had belched up from the depths of the mine. They blotted out the yard lights and gave the scene an ominous, surreal aura.

At the Ruddicks' house on lower Herrett Road, a stone's throw east of the pithead, no sooner had the tremors ceased than the three eldest daughters in the family rushed out the front door of their old company house. They raced across the field toward the mine as quickly as if they'd been shot from guns, following the same weedy path their dad had trod on his way to work about five hours earlier. When a tearful Norma Ruddick peered out the front window after her girls, she could see and hear the crowd at the pithead. Scores of figures were silhouetted against a glare of headlights in the bumper-to-bumper caravan of cars that was streaming down Main Street and into the DOSCO parking lot.

One vehicle that was not among those headed to the mine was Dr. Burden's old Jeep. Rather than rushing to the mine or

All Saints' Cottage Hospital, he'd made a beeline for the Main Street building that housed the DOSCO offices. It took him only a minute to make the short drive, but by the time he arrived there, several dozen men were clustered outside the door of mine manager George Calder's office; most of the men were miners who'd worked the day shift but were eager to do what they could to save any men trapped underground. As a study of the disaster would report, "Rescue work was regarded as a duty to friends, co-workers, and the occupation; it was part of the discipline of their work. Some had relatives in the mine. Others had themselves been saved in previous mine disasters. The majority had a general identification with their fellow miners by virtue of their common occupation and the knowledge that they themselves might face a similar situation in the future."[2]

An anxious crowd of miners, many of whom had worked the day shift at the mine, gathered at the pithead to await news on rescue operations and to volunteer to go below if needed. (Nova Scotia Archives)

Among the would-be rescuers was Cecil Colwell, who'd been on the afternoon shift but had gone home after suffering a cut arm in the seven o'clock mini-bump. He'd played an important role in rescuing men after the 1956 explosion at the now closed No. 4 mine, and he rushed to the DOSCO offices. He knew if any of his workmates had survived the Bump, they'd be praying for someone to rescue them. "They're probably all dead," Colwell's tearful wife had cautioned him. Cecil refused to listen or to believe her. He just couldn't.

Like all the men who rushed to the DOSCO headquarters, Colwell had no idea of the extent of the damage the Bump had caused. What he did know for sure was that no matter what, it might well be an hour, maybe even two, before the fully equipped rescue workers known as draegermen could marshal their resources, get to the pithead, and enter the mine to begin rescue operations. Time was a luxury the men trapped deep in the mine didn't have. That was why Colwell and the other would-be rescuers had hastened to the DOSCO offices to volunteer; all were willing to risk death by entering the mine "barefaced"—that is, without gas masks. Arnold Burden was ready to go with them. He'd worked in the mine during summers when he was a medical student, and he'd accompanied rescuers who'd gone into the No. 4 mine just minutes after the 1956 explosion. Burden knew what to expect. He understood the grave dangers posed by firedamp and by another bump. Both were very real possibilities.

All these concerns were swirling in the minds of George Calder and his assistant, Randolph Carter, as they repeatedly tried to reach one of the supervisors on duty at the mine. When someone finally answered the phone, it was the overman who'd been at the 7,800-foot level of the mine when the Bump hit. A breathless Jim McManaman—the uncle of Tommie Tabor's wife, Ruth—explained that the upheaval had knocked him unconscious. He'd just come to and was still struggling to clear his head. "We don't know what's happened," he said.

About all that McManaman could say for certain was that this upheaval had been a bad one. The air in the mine was so thick with coal dust that headlamps were useless, and firedamp levels were spiking dangerously in many areas of the mine. The overman added that he'd heard reports that many miners were injured and many others were presumed to be dead. However, as yet it was impossible to confirm any reports; the Bump had knocked out all the phone lines below the 7,800-foot level. This led McManaman to fear that many of the miners who'd been working deep in the mine were now dead, injured, or trapped deep underground. Calder shared those same fears and was already forming an action plan. Turning to Carter, he shouted, "Get me twenty men, fast."

That was easily done. No sooner did the assistant manager poke his head out the office door to call for volunteer rescuers than scores of hands shot up among the crowd. Almost every man there had a relative or friend who was working in the mine. Carter quickly chose a crew of volunteers and sent them over to the DOSCO storeroom to collect lights, shovels, axes, and any other tools they might need to dig their way through the rubble they knew would be clogging passageways in the mine. They would be going into the gas-filled mine barefaced, and doing so was to risk death.

While he waited for the volunteers to get ready for the planned rescue mission, George Calder called the mine again. This time, he ordered Jim McManaman to remove any coal cars from the tracks in the mine and replace them with man-rakes. Calder knew if the emergency rescue mission he was about to launch was going to have any chance of being successful, he'd need every trolley car he could muster to get everyone out of the mine as quickly as possible.

By 8:30 p.m., just twenty-four minutes after the Bump, Calder and his team of volunteers were ready to enter the mine. Much to Dr. Burden's dismay, he wasn't among them. Calder had advised him that because he wasn't a DOSCO employee or a government official, he'd need to wait for the draegermen to arrive. Then, if a mine company official was available to accompany him, he might be able to go down into the mine. Difficult though it was for Burden to accept that decision, he had no choice but to comply. Waiting wasn't something he was good at or that he enjoyed.

IN 1958, MINE disasters were big news, even more so than they are today. Many people burned coal to heat their homes. Industries were powered by coal, as were trains and ships. And so it was that coal mine cave-ins, bumps, and explosions generated headlines.

Within an hour of the Bump, a small army of journalists—both broadcast and print—from around the Maritime provinces and some from as far away as Toronto were en route to Cumberland County. All three of Toronto's daily newspapers—the *Telegram*, the *Daily Star*, and the *Globe and Mail*—sent reporters and photographers. So did several American newspapers including the *Boston Globe* and the *New York Times*, as well as the weekly news magazines *Time* and *Life* (even though it's a safe bet that the editors of these publications wouldn't have been able to locate Springhill or Cumberland County without consulting an atlas).

A seven-member contingent of journalists—three reporters and four photographers—from the *Halifax Chronicle-Herald*, the province's largest newspaper,[3] needed no such guidance. They knew the way to Springhill and were on the road and rushing to get there. Today, the drive by car from Halifax to Springhill takes two hours on a modern four-lane expressway. The situation was very different in October 1958. At that time, the seven members of the *Chronicle-Herald* team took three hours to reach Springhill, navigating what at that time was a narrow, twisty-turny two-lane road that could be hazardous at night. And sometimes deadly. Just a couple of hours after the *Chronicle-Herald* people made this drive, twenty-nine-year-old John Thompson, a reporter with the now defunct *Dartmouth Free Press*, lost control of his car while en route to Springhill and died in a crash near Truro.[4]

Walter ("Wally") Hayes was a decade younger than the ill-fated Mr. Thompson; however, he had similar career aspirations. Like any young journalist then, as now, Hayes was eager to cover a big story and hopefully in doing so make his mark in the newspaper world. In the autumn of 1958, Hayes was a fledgling part-time reporter-photographer—a "two-way man" in the journalism vernacular of the day—with the *Mail-Star*, the locally read afternoon little brother of the *Chronicle-Herald*, which was a morning publication with province-wide distribution. The youngest of three brothers, Hayes had little interest in school; that is, aside from the girls, the school pipe band, and the Army Cadet rifle team. He was in grade eleven when he dropped out and managed to land a job at the *Mail-Star*, first as a newsroom mail boy and then as a gofer and reporter-photographer trainee.

Hayes was earning a pittance, of course, but he was still living at home, and so he'd been able to scrape together enough cash to pay his share of the cost of a spiffy two-toned black-and-pink Studebaker, which he and his girlfriend co-owned, and to pay for burgers and fries at the popular restaurant they frequented. When he wasn't out with his girlfriend, Hayes hung out with his buddies in the gas station parking lot right across the street from his favourite burger joint, at the busy intersection of Cunard and Robie Streets. On the evening of October 23, that's where Hayes was. His car's radio was blaring a tune by Elvis when a voice suddenly interrupted with a news bulletin. A bump had rocked the No. 2 colliery at Springhill, the announcer said, and rescue efforts were under way to save men believed to be trapped deep underground.

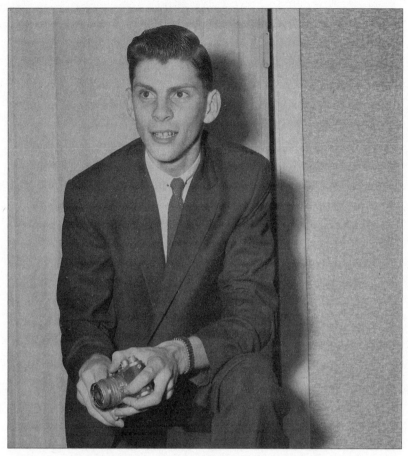

Eighteen-year-old Wally Hayes was the youngest member of the *Chronicle-Herald* editorial team in Springhill. (Courtesy of Wally Hayes)

Hayes immediately recognized that this would be a major news story, and so he hopped into his car and sped downtown to the *Chronicle-Herald* newsroom on Argyle Street. The place was buzzing with activity by the time he got there. Managing editor Frank Doyle was on the phone, dispatching members of the previously mentioned team of reporters and photographers that went speeding off to Springhill; Hayes dearly would have loved to be one of them, but he knew this wasn't going to happen. At age eighteen, he was too junior for such a plum assignment. What's more, apart from his photographic skills, he had no special talents that recommended him for inclusion. He'd have been the first to tell you he wasn't a strong writer. No matter. As it turned out, it was precisely his lack of special credentials that proved to be his key selling point when early next morning he steeled up his nerves, marched into Frank Doyle's office, and announced that, if possible, he wanted to go to Springhill. His rationale was simple: he could serve as a backup and as a gofer for the *Chronicle-Herald* reporters and photographers when they were busy covering breaking news. To Hayes's surprise and great delight, the usually gruff managing editor agreed with this cheeky request. "Go," said Doyle.

"My job was not to take pictures," Hayes would recall many years later, "but get film from our photographers . . . then hotfoot it to the darkroom in Amherst, process and print negatives and [then] get the prints to the CP [Canadian Press] Wirephoto as quickly as possible."[5]

What made the otherwise mundane opportunity exciting in Hayes's mind was that the trip to Springhill was his first out-of-town assignment as a *Herald* employee. The experience was an eye-opener for him, one he'd never forget. "It was late October. Many of the leaves were off the trees, and the sky was grey as I drove to Springhill. To be honest, the town looked kind of grubby to me," Hayes would recall many years later. "It reminded me of one of those frontier towns you used to see in western movies. There were some buildings with false facades."[6]

Adding to the novelty of Hayes's experience was the fact that because there was no hotel in Springhill, the *Chronicle-Herald* staffers were bunking in a small hotel midway between the town and Amherst, 16 miles to the northwest. The eight-member team shared a single room. This was not only uncomfortable but also problematic since one of the reporters was Mary Casey, a senior journalist who normally covered provincial politics. The fact that Casey was a member of the *Chronicle-Herald*'s news team in Springhill served to further underscore how important the newspaper's senior manager deemed the Bump story to be.[7]

Whenever Casey used the hotel room, propriety dictated that the male reporters and photographers had to sleep in their unheated cars. And on a couple of chilly nights, Wally Hayes would remember sacking out in his Studebaker or on the floor under a desk at the DOSCO offices. "And if not for the fact the Salvation Army had set up a tent and a field kitchen, I don't know where I'd have eaten," he said.

Hayes and the rest of the *Chronicle-Herald* news team endured their ordeals, as did the scores of other out-of-town journalists who'd descended upon Springhill to report on the mine disaster and on rescue efforts. Uncomfortable though they doubtless were, their travails and discomforts paled when compared with the traumas the people of Springhill were suffering through or the nightmarish, life-and-death ordeals of the Bump survivors, who now found themselves entombed deep underground with little food and water, praying for a miracle rescue.

CHAPTER 13

THE TWELVE

Thursday, October 23, 1958
8:30 p.m.

A FEW MOMENTS IS ALL IT TAKES FOR A LIFE TO CHANGE irrevocably. And that's how it was for Harold Brine. His life, and that of his wife and daughter—and indeed of every person in Springhill—changed in the fleeting instant it took for the Bump to shatter the No. 2 mine.

The amount of energy released when the earth heaved was so great that Brine had no time to be afraid, much less to cry out. The force of the tremor sent him flying like a bag of feathers. At the same time, a loose chunk of wood smacked him on the head, knocked him senseless, and imprinted a permanent bruise on his left ear.

The next thing Brine knew, he was flat on his back in a pile of mine waste. He had no idea how he'd gotten there, or of how long he'd been there or how long he'd been unconscious. His headlamp, which miraculously was still shining, barely penetrated the thick storm cloud of coal dust that swirled in the air above him. Brine felt groggy and was having trouble catching his breath, and so for several minutes he lay there semi-conscious, struggling to decide whether he was alive or dead. He wasn't exactly sure. Only gradually did it dawn on him that yes, he *was* still alive, and that there'd been a bump—the Big One that everyone had so long dreaded.

Somewhere in the distance, Brine could hear bits of coal and rocks falling. From time to time, this muted din was punctuated by what sounded like someone moaning in pain. Or could it be mine timbers shifting and creaking? It was all too indistinct, too far away, for Brine to be certain. His thoughts drifted homeward. He thought of wife, Joan, his beloved two-year-old daughter, Bonnie, and the house he was building; he still had no end of work to do before the place would be done. He wondered if he'd ever get the chance to finish it.

When Brine's mental fog finally began to clear, he sat up and looked around. As the beam of his headlamp stabbed through the darkness, the scene it revealed was stunning. The area in which he and his co-workers had been working mere minutes before was now clogged with a mass of debris—miners' tools, shattered helmets, and an avalanche of large clumps of coal that had come tumbling down from the coal face above.

All this devastation had happened when the mine floor heaved upward. It had done so with such ferocity that it crushed many of the timber packs supporting the roof; they'd snapped like matchsticks. There was now less than six feet of headroom in the space in which Brine found himself trapped. In some spots, the floor was within three feet of the roof; in others, it was even closer. And as he peered through the darkness and the clouds of coal dust, Brine felt as though he'd been kicked in the stomach. The sight of the crumpled, lifeless body of a man he'd known and worked with for several years left him feeling sickened and weak at the knees. In that instant, it dawned on him how intense the Bump had been.

Brine had worked as a draegerman in 1956, helping to rescue miners who found themselves trapped deep underground in the No. 4 mine. Now, to his dismay, he realized he was the one who needed to be saved. When that sobering awareness clicked into place in his mind, it occurred to Brine that the strange noises he'd been struggling to identify were the cries of co-workers. That realization prompted him to clamber to his feet. He felt woozy. His ribs hurt, and his head ached. Despite his discomfort, he was relieved to find he could walk, albeit unsteadily.

The uneven surface beneath his feet sloped off in a westerly direction, just as the main and back slopes did. In the perpetual darkness of the coal mine, this made for treacherous footing at the best of times; a man had to tread carefully. The upheaval had made the hazards infinitely more treacherous.

After taking a few moments to calm himself and get his bearings, Brine began picking his way through the rubble. His progress was painfully slow. He'd gone only a short distance when his headlamp found someone lying on the mine floor. Although the man's face was a mask of coal dust, Brine somehow recognized Joe Holloway's angular features. Joe, a soft-spoken thirty-five-year-old, had been wounded while serving overseas with the Canadian Army during the war. He was Hughie Guthro's pal. The two men had a special bond after having survived the 1956 explosion together. Brine knew that while Holloway seldom complained, he'd never been shy about saying how much he hated working in the No. 2 mine and had been adamant that the place was "an accident waiting to happen." As it had turned out, he'd been horribly right in that assessment.

Holloway was half-buried under an avalanche of loose coal that had rained down on him. After Harold Brine had dug him out and revived him, the two men went looking for other survivors. By now, Hughie Guthro had lapsed into unconsciousness and fallen silent. It was Joe Holloway who chanced to find him. He did so after stumbling over what he initially took to be a rock or a chunk of wood. Holloway was stunned to discover the obstacle he'd encountered was Hughie Guthro's head. That was the only part of his body that was sticking out of a coal pile.

Once his shock at finding what looked like a bodyless head had abated, Holloway realized that Guthro was still alive. Dropping to his knees, he roused his pal and then began clawing at the coal with his bare hands to dig him out. Moments later, Brine arrived on the scene and joined in the rescue. The two of them were digging frantically when suddenly they both began to feel light-headed. Being experienced miners, they realized the air near the roof of the mine was heavy with firedamp, and they were on the verge of passing out. The gas, being lighter than air, rises in a mine, just as smoke does in a burning building.

Under normal working conditions, the mine's ventilation fans kept the level of this methane-laden hazard in check. To make sure that was the case, the underground managers also carried lanterns, which they hung from the roof of the mine at the top end of each coal face. By monitoring the intensity of the lantern flames, the managers could keep track of how much firedamp was present. This was the latter-day version of the proverbial "canary in the coal mine." Air normally contains about twenty-one percent oxygen, and fire requires at least sixteen percent oxygen content to burn. If ever the oxygen level in a mine dropped below that magic sixteen percent figure, work was paused, and the miners dropped down to the mine floor until the air cleared. If it didn't clear, they made their way to the surface as quickly as possible.

Undeterred by the presence of firedamp, Brine and Holloway flopped down to the floor and waited to regain their senses. Only when their heads had cleared did they resume digging, this time with extra haste. "Hang on, Hughie," Brine assured Guthro. "We'll get you out of there. If you're going to go, we're going with you."[1]

Both Brine and Holloway knew that if either of them had been the one who was trapped, Guthro—or any other miner— would have come to their rescue or would have died trying to do so. That's how it was in the mine: all for one, and one for all. That informal, implicit agreement was an extension of the same cooperative spirit that saw miners look out for each other, much in the same way they scrubbed each other's backs in the wash-house shower room. In the harsh, unforgiving conditions when men are labouring in a coal mine deep underground, survival often depended on such cooperation.

A miners' safety lamp of the sort used in the Springhill mine. (Ken Cuthbertson)

ONCE HAROLD BRINE and Joe Holloway had freed Hughie Guthro, the three men lay sprawled on the mine floor for several minutes. The thick clouds of coal dust that continued to foul the air clogged their noses and singed their lungs. The only positive was that the air was breathable. That being so, as the trio slowly regained their senses, they began to survey their surroundings. What they saw hit them like a gut punch.

The cavity in which they were trapped was about as big as a typical Canadian living room—roughly 20 feet long by 15 feet wide. With barely enough headroom in which to stand upright and an uneven, sloping floor, the space was no longer as accommodating as it had been before the Bump. Compounding the men's problems was the awareness that the batteries of their headlamps, which were good for about ten hours, had only three or four hours of life left, at most. When the lights died, they'd be in a darkness so total you could almost touch it, feel it.

While the prospect of being stranded in the dark terrified some men, it didn't bother Harold Brine. Not much anyway. He'd never been afraid of the dark when he was a boy, and he wasn't afraid of it now. After seven years as a coal miner, he'd grown accustomed to working in the pitch black with only the beam from his headlamp to guide him. "Those headlamps gave off a pretty good light," he said. "And as far as being in the dark, you get used to it. It's like when you're in your own home if you know where everything is."[2]

Brine and his companions were still figuring out their own situation when Ted Michniak began hollering. He was somewhere out there in the darkness, as was someone else whom Harold Brine thought sounded like Gorley Kempt. But Brine was puzzled. How could that voice belong to Kempt? he wondered. Gorley had been working at the top of the 13,000-foot coal face; that's where he was when Brine had last seen him. However, unbeknownst to Brine, when the Bump hit, Kempt had been making his way down to the level below to deliver a wrench to the men trying to restart the engine that drove the coal pans. "Suddenly, the [floor] seemed to explode," said Kempt. "Everything flew with a terrible rushing of noise and wind."[3] Then there was only darkness and a great nothingness. The force of the upheaval had sent Kempt hurtling through the air. When he regained consciousness, his feet were up; his head was down. Fortunately for him, while he was bruised and badly shaken, he was otherwise unhurt. For that, he could only thank heaven.

Surprisingly, Gorley Kempt's headlamp was still working, and when he turned it on, its beam revealed a frightening scene: he saw a mass of rocks, dirt, tools, shattered helmets, crumpled coal pans, entire mine cars, lengths of rail, chunks of timber, pieces of pipes, and snarled wires. In short, everything that at the start of the afternoon shift had been the backdrop and equipment for work in the No. 2 mine was now jammed up against the roof, crushed almost flat.

Even more unnerving was the gut-wrenching sight of bits of the bodies of a couple of grotesquely mangled miners. A bloodied arm, a hand, a foot, and what looked like it had once been a man's head protruded from the compacted roof. The scene had a horrifying, nightmarish air of unreality to it. Small wonder that once Kempt extricated himself from the tangle and got his feet under him, he scrambled out of there as quickly as he could. After several anxious minutes of threading his way through the debris maze, he emerged into the chamber where Brine, Holloway, and Guthro were huddled together. Cutting through the darkness, to Kempt's eyes their lights were beacons of hope.

It wasn't long before other disaster survivors began appearing, seemingly out of nowhere. Eight more men did so. Several who emerged were injured. Levi Milley, for one. The Bump had knocked him so hard he was senseless. The shattered mine was eerily quiet when he came to; how long he'd been unconscious, he had no idea. All he knew for certain was that he was alone and was in pain.

It was while he was desperately looking around for a way out of there that his heart skipped a beat. A few feet to his left, Milley's light flashed on a pair of eyes glinting white in the dusty darkness. The man's headlamp was nowhere to be seen, and his face was so thick with coal dust that Milley couldn't identify him. The prostrate man was sobbing and appeared to be totally out of it.

"Who are you?" Milley asked.

"It's Caleb Rushton," came the muted reply. "Help me. Oh, please help me. Something is pinning my leg, and it hurts like the dickens."[4]

It took no small effort for Milley to lift the huge rock. He was able to do so just enough that Rushton was able to wriggle free. Apart from a painful gash on his head, he'd been fortunate. Rushton hadn't sustained any serious injuries. If there'd been more headroom, he might have been able to walk. However, in the confined space, he and Milley could only scuttle crab-like, inching forward on hands and knees in the only direction they could go. And they hadn't gone far when they heard the chilling sound of someone moaning. When they searched around with their headlamps, they spotlighted a small crevice. Inside that opening was an injured man. He was lying with one of his legs pinned under him at a strange angle.

It was Joe McDonald, a thirty-eight-year seasoned veteran of minework. Another survivor of the 1956 explosion, he'd been working at the 13,400-foot level when the earth moved. McDonald had come through the horrors of that earlier tragedy unscarred physically, albeit with hair that was prematurely grey. Mentally, he hadn't fared nearly as well. McDonald had endured a difficult, emotionally harrowing childhood, and as such he'd always been deathly afraid of the dark. Not surprisingly, his experiences being trapped underground had reignited and accentuated his fears. For many months after his rescue, he'd slept each night with a bedroom light on.

McDonald's angst was compounded by his heavy drinking. He'd been an alcoholic most of his adult life. The trauma of his 1956 ordeal had increased his dependency on booze and accentuated his woes. In desperation, he'd finally joined Alcoholics Anonymous and regained control of his life. Despite having done so, McDonald still hated being underground and was deathly afraid of the bumps that had been rocking the No. 2 mine with such alarming frequency of late. He often said he'd gladly have quit the mine and earned his living another way if finding a job in rural Nova Scotia at this time hadn't been harder than finding a five-legged horse. As a result, while he groused a lot, he hadn't quit the mine.

Like Joe Holloway, McDonald saw all of his worst fears become real the day the Big One happened. He'd been kneeling with one leg under him when the floor heaved and sent a section of the coal pans flying. It came crashing down atop his left leg. "It just seemed as if the coal face came towards me," he would recall. "I lost my light and my cap. . . . I didn't feel my leg until I tried to walk."[5]

That was just as well, for McDonald's injury was horrendous. When he tried to stand, his leg buckled, for it was broken in three places. Not knowing this, Levi Milley grabbed hold of McDonald's contorted leg and tried to straighten it. The off-the-charts pain this caused McDonald set him shrieking. In that same millisecond, Milley recoiled, reflexively withdrawing his hand. He was horrified to realize he'd seized on a shattered thigh bone that protruded through McDonald's skin. Small wonder McDonald had cried out as he had or that he pleaded, "For God's sake, don't leave me!"

Milley and Rushton did what they could to comfort McDonald. They erected makeshift props around him in hopes of preventing more coal or other debris from falling on him. They also assured him they'd stay there with him. However, once he lapsed into unconsciousness, Milley and Rushton decided their wisest course of action was to continue looking for a way out. Hopefully, if they could find one, a rescue crew would then be able to extricate McDonald and rush him to the hospital. Like the other men who'd survived the Bump, Milley and Rushton initially had no idea of the extent or severity of the damage to No. 2 mine. They knew only that where they'd been working, the situation was bad. The mine was in a shambles, and many of their co-workers were dead. The urgency and uncertainties of the emergency were underscored when Ted Michniak, accompanied by a shower of pebbles and coal, suddenly came sliding down the incline and into view. He was a frightful sight.

Michniak was a big, muscular man. When he had regained consciousness after the Bump, he'd been bleeding from the ears, and he also had suffered a broken shoulder and a shattered left wrist. His pain was excruciating. After screaming himself hoarse, Michniak finally calmed down and gradually was able to regain his composure. When he did, he somehow dug himself out from under a fall of coal that had pinned him. Then, despite being delirious with pain and having lost his bearings, he was determined to get out of the mine alive. If that wasn't possible, he wanted to find someone else, another survivor. Michniak didn't want to die, but if he did, he didn't want to die alone in the dark. When he heard voices, Michniak instinctively began to claw his way toward them. It was while he was doing so that he encountered Milley and Rushton.

The two men suggested that Michniak stay with Joe McDonald while they went off to look for a way out. That's when they bumped into Bowman Maddison. He'd been lucky beyond measure. Maddison had narrowly escaped death when the Bump threw him for a loop at the same moment that a huge chunk of coal hurtled past his face and grazed his nose. That same projectile severed the cord that connected his cap lamp and the battery on his belt. "I never felt no shock or anything [*sic*]. It just cut my light right off, just like somebody had cut it with a knife. . . . I think the bottom came up. I don't know how long I was [knocked] out."[6]

Entombed deep underground in a collapsed mine with no light and no sense of the chaos all around him, Maddison had almost no hope of survival; the odds would have been slim to non-existent. If Levi Milley and Caleb Rushton hadn't happened along, he surely would have suffered a slow, torturous death, buried alive.

Maddison eagerly joined his rescuers in what turned out to be a fruitless search for a way out of the shattered mine. The men were trapped. What was even worse, they'd learned that with the pit in a shambles and full of firedamp, escaping the No. 2 mine wouldn't be easy or quick. Their only real hope was that a rescue crew could reach them before there was another bump or they died for lack of food and water. Regardless, they wouldn't last long.

Larry Leadbetter had come to that same conclusion. Leadbetter, who was just twenty-two and a father of two young children, was one of the youngest members of the mining crew who had been working on the coal face at the 13,400 level. "It seemed like a little shiver under your feet, and then a sudden upheaval, and that was it. I think . . . I must have hit my head on a pack or something. I can't recall how long I was out. It might have been a day, or an hour, or it could have been five minutes. I really don't know."[7]

However long it was, when he came to, Leadbetter was disoriented and shaky. He knew there'd been a bump, and it had been a bad one. The evidence stared him in the eye when he turned on his headlamp. He found he was lying in a small space, with the roof no more than 10 inches above him. There wasn't even enough room for him to sit up.

Fear had paralyzed Leadbetter. There was just one thought in his head: he had to get out of there. He would have no memory of doing so, yet somehow, he scrambled out of his would-be tomb. The effort to do so left him feeling spent and totally disoriented. Even worse, he had no idea which way to go next. He was done.

He collapsed into a heap and lay there sobbing as he waited for whatever fate had in store for him. As it happened, it was not death that found him, but rather Levi Milley, Caleb Rushton, and Bowman Maddison. When the three men appeared, Leadbetter suddenly popped up as quickly as a jack-in-the-box. "Oh, my God! My God! Help me. I don't want to die," he bawled. "Don't leave me here alone!"

"Jeezus . . . you scared the crap out of us!" said Levi Milley.

Bowman Maddison was also taken aback, and Leadbetter's frantic outburst had startled him. His temper flared. The stress of the situation was already starting to get to all of them. "None of us want to die, damn it!" he screamed.

Fortunately, Levi Milley somehow had kept his head. He reassured Leadbetter they wouldn't leave him and managed to defuse Maddison's anger. When he'd done so, the four men fell silent while they mulled over their predicament. They were still doing so when they heard voices calling out. It was Harold Brine, Joe Holloway, Hughie Guthro, and Gorley Kempt. The foursome had heard Maddison and Leadbetter yelling, and now they were shouting in hopes they'd attract the attention of what they prayed were rescuers who'd come to save them.

The outcry set Milley, Rushton, Maddison, and Leadbetter picking their way downward along the fractured coal face until they reached the spot where their four co-workers had found refuge. It wasn't far, but the going was slow.

There were now ten survivors, and soon two more emerged out of the darkness. It seemed to Brine that Ed Lowther and Fred Hunter had appeared suddenly and unexpectedly, as if they'd been conjured up by a magician. "I never knew where all the guys came from down there," said Harold Brine. "There were just four of us to start, and then it ended up that there were twelve in total."[8]

Ed Lowther, a forty-five-year-old WWII army veteran, had worked as a miner since 1930. Hunter, who was forty-nine, had done so for thirty-one years. He was a quiet, unemotional man, and there was a good reason. Hunter had spent nine months in a body cast after suffering a broken back in a 1952 workplace accident, and the 1956 killer explosion had trapped him underground for four days. And now his run of bad luck was continuing. Strike three.

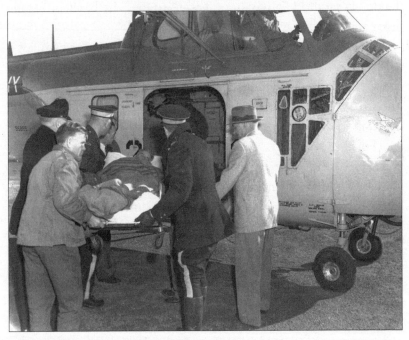

Military helicopters were on standby and ready to fly to Halifax any miners who were too severely injured to be treated in Springhill. (Nova Scotia Archives)

The Bump had KO'd Fred Hunter. When he came to, he was horrified to see that his left leg was badly bruised and the internal bleeding was causing it to swell quickly. A piece of the mine machinery had dealt Hunter a glancing blow as one or maybe both flew through the air. Hunter was in a world of pain as he groped his way along the coal face in search of a way out. While doing so, he came upon a trio of dead men, one of whom was jammed up and hanging from the roof of the mine. In the dim light, Fred Hunter couldn't be certain, but he felt sure the corpse he saw was that of his twin brother Frank. The mere sight of the body left Hunter feeling nauseous. He was barely aware of having spoken up to identify himself when Ed Lowther began doing a head count and assessing everyone's condition; Lowther's WWII military training was kicking in. "None of us are seriously hurt," he announced. As rosy as that assessment of the situation was, it wasn't all that accurate.

It was true there were a dozen men alive at the top end of the coal face on the 13,400 level. However, two of the survivors were seriously injured, and the grim reality was that all the men were clinging to life by the flimsiest of threads. Harold Brine, Gorley Kempt, and one or two of the other men carried Joe McDonald and Ted Michniak into the space where the group had gathered. They were doing their utmost to make the injured men comfortable. The most pressing question that lingered in the mind of every survivor was a simple one: could they—and would they—survive until rescuers arrived? That is, *if* rescuers were trying to reach them.

CHAPTER 14

THE SEVEN

THURSDAY, OCTOBER 23, 1958
8:40 P.M.

SO CLOSE AND YET SO FAR. FOUR HUNDRED FEET OF DIRT,
sandstone, and coal separated two groups of miners who found
themselves trapped in the No. 2 mine. The distance between the
two groups of men could have been a million miles; it didn't make
a whit of difference. They had no way of knowing who else—if
anyone—had survived the Bump, and if anyone else had survived,
where those men might be in the mine. Despite this uncertainty,
all the entombed men had one thing in common: time was their
mortal enemy.

Harold Brine and eleven co-workers who found themselves
trapped at the 13,400-foot level understood their predicament. And
400 feet away, up near the top of the same coal face, so too did
Maurice Ruddick and his companions; yet for one of them—the
ill-fated Percy Rector—time had ceased to have any meaning. He
was imprisoned in a real-life nightmare. Caught in an upright

156

position with his right arm crushed between the unyielding, immovable timbers of a collapsed pack, for Rector there was no relief from his agony. It was excruciating, and it was constant. His only respite came as a result of the toxic gas that was now pervasive in the mine.

When the earth moved, the firedamp that filled the air rendered many of the men unconscious. "The gas [that had] started coming in, seeping in on us, put us to sleep," Herb Pepperdine would recall in a television interview a half-century later. "I don't know how long I slept, but when I woke, I thought it was in my own bed."[1]

Maurice Ruddick's experience was markedly different. He knew he wasn't at home in his own bed. Buried in coal up to his waist, he awoke unable to move his legs. A black deluge had rained down upon him, pinning him to the mine floor, dislodging his safety helmet, and temporarily knocking out his light. Fortunately for Ruddick, his arms remained free, and when he flicked the switch on his headlamp, it still worked.

After digging himself out, although he was as shaky as a toddler taking first steps, Ruddick staggered to his feet. In the confined space, his helmet scraped the roof. Then in the next instant, he felt woozy. Firedamp was filling the space nearest the roof. Ruddick crumpled in a heap when he breathed in the toxic gas. He lay on the mine floor in a semi-conscious state for several minutes. When next he was aware, it was because he was startled by the sound of a man crying out. A glance at the dial of the luminous wristwatch his wife, Norma, had given him told Ruddick it was 8:40 p.m. Thirty-four minutes had passed since the Bump.

Once his mind cleared, Ruddick was able to take stock of his situation. About all he was able to discern was that there'd been a major upheaval. He still had no idea of exactly how bad it had been. At the same time, he was puzzled. In the distance, he could hear someone calling out, pleading for help. Twenty minutes later, after picking his way through the debris that now all but clogged the walkway along the coal face, Ruddick discovered who had been hollering and why.

One look at Percy Rector was all it took for Ruddick to know the bump that had happened was the Big One he and everyone else had been dreading. Ruddick knew there was nothing he could do for poor Percy; if a rescue crew didn't come soon, Rector was done for. And even if a rescue crew did reach them, the only way they'd ever be able to free him was to amputate his arm and then pray they could somehow staunch the bleeding long enough to carry him to the surface. Any such ordeal would be gruesome beyond measure for both Rector and whoever wielded the saw.

Cutting through the skin, muscle, and ligaments of a living human being isn't easy, even if the knife or saw is razor-sharp. And the instant the first incision is made, bleeding becomes the overriding concern. The only way to stem the torrent is to fashion a tourniquet of cloth, rope, or some other suitable material. Regardless, time is of the essence.

During the American Civil War, working under the most primitive conditions imaginable—but still better than those you'd find deep underground in the darkness of a crumpled coal mine— battlefield surgeons amputated the limbs of some thirty thousand soldiers. During these procedures, a scalpel was used to cut through skin and then a double-bladed Catlin knife for muscle and tendon. Next, to finish the job of severing a limb, the surgeon employed a bone saw—the tool that gave rise to the slang term "sawbones" for a surgeon. None of this was pretty, nor was it neat, especially when there was no anesthetic to ease the amputee's pain. The closer the amputation was to the chest, the greater the chance of death as the result of blood loss or other complications. What's surprising about all this is that the survival rate for soldiers who lost a limb to amputation was still almost seventy-five percent.[2]

None of the miners who were trapped with Percy Rector knew any of this, mind you. How would they? They weren't doctors, nor did any of them have any medical training. Understandably, they were all loath to even consider the prospect of severing Rector's arm.

It was plain-talking, no-nonsense Herb Pepperdine who was the first to do what others doubtless were thinking. Ruddick watched Pepperdine retrieve the blade of a well-used pit saw protruding from a pack that had been under construction when the Bump hit. Because the saw's handle had broken off, Pepperdine could retrieve only the blade. When he did so, he concealed it behind his back in case Rector was watching. But there wasn't much chance of that— he was delirious and slipping into and out of consciousness.

Stepping close, Ruddick whispered to Pepperdine that he hoped they wouldn't have to use the saw. Pepperdine nodded in agreement. "I got it just in case," he said.

Both men feared that any effort to amputate Rector's arm would kill him. Neither of them wanted to find out if that indeed would happen. Better, they reasoned, to wait for the rescuers they prayed were coming. Those men would be equipped with better tools, more know-how, and stronger stomachs. With any luck, Dr. Burden or another medical man might even be with them.

Mercifully, the decision about what help, if any, to offer Rector was deferred for now. At that moment, he was comatose, and an ominous silence enveloped the rest of the men.

AT ABOUT 9:30 P.M., the calm was shattered by the arrival in the chamber of another survivor. Out of the darkness, Frank Hunter appeared. He came sliding down the slope accompanied by a deluge of loose coal.

The Bump had sent a huge rock bouncing off Hunter's helmet. The blow had stunned him, causing a mild concussion, and ruptured his eardrums; a trickle of blood ran from his ears. The same rock that had felled him had also landed a crushing blow to Hunter's thigh, although the resulting severe internal bleeding wasn't initially apparent. Like so many other men, he'd been knocked out, and his pain was temporarily eased when he passed out after having breathed firedamp. He reckoned he'd been unconscious for almost ninety minutes.

Frank Hunter was the twin brother of Fred Hunter, one of the dozen miners trapped one level below. Fred was certain his brother was dead, but he'd been mistaken. Frank knew nothing about this, of course, and that was just as well. Although his leg was severely injured, for now at least he was still very much alive. And he was happy to be so.

With the arrival of Frank Hunter and Doug Jewkes, a wiry thirty-seven-year-old who'd suddenly appeared at the top end of the 13,000-foot coal face, the number of miners who were huddled together there had increased to seven, counting Percy Rector. The chamber they occupied was sizable—about 40 feet long by 12 feet wide. Optimists in the group were thankful there was enough headroom for them to stand up. The pessimists only saw this chamber as the dark place where they were going to die, and if the mine were closed and sealed off, it was where they might be forever entombed. "I have a piece of chalk in my pocket," said Garnet Clarke. "If you all call out your names, I'll write them down on the timbers."

That suggestion, simple and pragmatic though it was, didn't sit well with all the men; from the darkness came an unidentified voice wondering aloud why they should bother to do this. Ignoring the question, Clarke scribed his own name on a timber and then added the names of the other five men, all of whom identified themselves. "You can write down my name," said Maurice Ruddick. "I don't care what anyone says. I'm not going to die here, and neither are any of you."

Ruddick, being a man of deep and unshakable faith, believed with all his heart that he'd see his wife, Norma, and his kids again. He remained confident that, God willing, he and his mates would be saved—even Percy Rector. However, those notions were a tough sell under the circumstances, here in the pitch black, far below the surface of the earth. It didn't help that periodic bursts of firedamp continued to render the men unconscious for brief intervals or that they had so very little food and water. A couple of stale sandwiches and half a canteen of water weren't enough for one man, let alone seven. Most of them had eaten at least half the food in their lunch boxes during the six o'clock break. They'd also emptied most of the water from their canteens. No one had made any effort to conserve food or water. They'd had no reason to do so. Now in the wake of the Bump, they were trapped, and unless there was a miracle rescue, odds were high they'd all die in the mine.

Currie Smith, Garnet Clarke, and Herb Pepperdine had returned disappointed and discouraged after having worked their way to the bottom end of the coal face in search of an escape route. Clarke had no more success when he'd gone out with Ruddick looking in another direction. They crawled about 60 feet along the 12,600-foot level, which was above where they'd been working. It was slow going—and all for naught in the end. "When we got back down the wall we started rooting around for odds and ends of water," said Ruddick. "We had six or seven bottles at the last of it, but some of them had very little water in them or at best they were half full. All the lunch cans but one had been smashed. The worst of it was that we burned out most of our lamps getting out of the level and most of our energy, too."[3]

Undeterred, Ruddick went off on his own to explore one more time. He returned to the spot at the top of the coal face where he'd been immediately after the Bump. His efforts to find a way out there were as unsuccessful as his earlier foray with Garnet Clarke had been. The mine floor in the area where Ruddick crawled on his stomach was wedged up against the roof so tightly that a bug couldn't have crawled through the tiny opening. Ruddick had only gone about 60 feet along the wall when he realized, to his horror, there wasn't even enough room for him to turn around when he decided to retreat. It took him an hour to do so, wriggling backward inch by inch. The effort was torturous and left his hands and elbows bloodied and raw. After finally returning to the chamber where the other men were waiting, a sadder but wiser Ruddick reached for the Aspirin bottle he always carried in the pocket of his work pants. As he was doing so, he had a sudden idea. "I've got some Aspirins. Has anybody else got any?" he called out.

A quick search by the other men turned up five more tablets. Combined with the ones Ruddick had in the bottle in his pocket, there were about a dozen tablets in total. Hearing Percy Rector's pitiful cries had prompted Ruddick to realize that Rector's need for pain relief was far greater and more urgent than his own. It wasn't much, but when Ruddick placed three tablets in Rector's mouth and held a water bottle up to his lips, Rector managed to gulp everything down. With his pain now dulled, even if only marginally, he slipped into unconsciousness once again.

In the silence that followed, the other men had time to collect their thoughts. Their minds drifted to home and to their loved ones. To save the batteries in the headlamps, they all switched off their lights, for the bulbs were already starting to grow dim. Once the lights died, the men knew they'd be in the dark for the duration. Sitting there now, the talk soon turned to how soon rescuers might come looking for them. "What if they don't come?" someone asked.

That was a possibility no one wanted to discuss, even though they all knew that possibility was a very real one. If the firedamp was thick or conditions were otherwise too unsafe for the rescue workers, DOSCO management might choose to close the mine and wait for the underground hazards to abate, even if that took weeks. If that happened, any rescue mission would become a mission to recover bodies.

That distressing thought set Maurice Ruddick thinking. He wondered how his wife could ever hope to raise their dozen kids if he died here in the mine. He prayed that Norma wouldn't have to face that possibility. Even with the meagre life insurance policy he paid premiums on each month and any survivor benefits DOSCO provided, it would be tough going for her and the kids. That thought was troubling, and so as he often did when he felt stressed, Ruddick sought comfort in music. He began to sing an old hymn that had come to mind. He sang softly and gently as he often did at home when it was late at night, all the kids were asleep, and the house was quiet. Ruddick would sing for himself. And sometimes he sang when he was nestled in bed next to Norma. Singing brought him peace of mind and comfort. His intonations were like chanting; they had the same soothing effect on the others. Then out of the darkness came a plea: "Maurice, can you sing louder so we can all hear you?"

Ruddick did as he'd been asked. "I come to the Lord in prayer / Though my path is narrow . . ." he sang, his rich baritone filling the chamber. However, it wasn't long before his voice trailed off. For some reason even Ruddick couldn't explain, he didn't feel much like singing this night. He fell silent, as did the other men. Each of them was now alone in the darkness. Each man had his own prayers and his thoughts of home and the loved ones he might never see again. Suddenly, it seemed frighteningly clear to Ruddick and to every one of the men trapped with him that they'd all likely meet their ends here in this miserable place.

CHAPTER 15

TO THE RESCUE

———

Thursday, October 23, 1958
8:40 P.M.

In the minutes immediately after the Bump, great billowing clouds of coal dust poured forth from the deepest recesses of the No. 2 mine. The eerie pall that descended on the pithead matched the growing despair of the crowd that rushed to the mine hoping to learn the fate of loved ones. The scene was surreal, a replay of the turmoil that roiled the town in the wake of that November 1956 killer explosion.

On another tragic night two years later, "Hundreds of numbed townspeople [now] waited in the October cold for word of brothers, sons, and husbands trapped in the deeps," as one observer noted.[1] Dayshift miner Paul Melanson, whose brother Leon was among the scores of men missing down below, articulated what many people in the subdued crowd were thinking. "[The outlook] is pretty grim," he sighed.[2]

Mine manager George Calder felt the same way, and he felt a sense of personal obligation to do whatever he could to rescue any men who'd survived the Bump. And so, he moved quickly and decisively. It was barely half an hour after the upheaval, clouds of coal dust still hanging in the air, when rescue efforts at the mine got under way. The people who'd gathered at the pithead fell silent when they saw Calder and a crew of stony-faced rescue workers arrive. Among them were union local president Monson Harrison, Nova Scotia deputy minister of mines Allan Fuller, and seventeen men chosen from among the scores of miners who'd volunteered to enter the mine barefaced; that is, with no breathing apparatus.

The usual practice when a serious bump or cave-in occurred was for DOSCO managers to await the arrival of draegermen with their specialized gear—the scuba-like breathing apparatus they wore on their backs—that enabled them to venture into areas of a mine where firedamp was at toxic levels. However, Calder knew time was of the essence today; every second counted. The faster rescuers could reach any injured or trapped men, the greater the chance that lives could be saved. Calder feared it could be an hour or even more before draegermen were ready to enter the shattered mine. Some of these specially trained rescuers were locals; others were en route to Springhill from Nova Scotia DOSCO mines at Stellarton, Glace Bay, New Waterford, and Sydney Mines.

It was extremely dangerous to go underground barefaced after a bump or an explosion in a coal mine. Calder and the men who'd volunteered to accompany him on his rescue mission knew this all too well, yet they didn't hesitate to take the risk. Not for a moment. They and every other miner implicitly subscribed to "the miners' code," which was a timeless, universal, and unwritten canon with two guiding principles. The first was always to believe, no matter what the circumstances were, that a man who's missing is still alive. The second was that a miner is obliged to do everything he can to bring to the surface—alive or dead—any fellow miner who has been trapped underground.

These principles were doubtless top of mind for the nineteen men in the advance rescue crew with George Calder when they took their seats in a man-rake that sat waiting for them at the top of the main slope. It was 8:55 p.m. by the time everyone was seated and ready to go. After taking a last quick look around, Calder waved to the head-drive winch operator, and the big drum that drove the trolley system slowly began to turn. As it did, the trolley car carrying the rescue crew lurched forward and began its descent into the depths. It took less than half a minute for the workers at the pithead to lose sight of the rake and the rescuers' headlamps. Both were quickly swallowed up by the perpetual darkness of the mine. The onlookers who stood vigil at the pithead could only wait and wonder what horrors the men in the rescue crew might encounter. Would they all return to the surface alive?

George Calder had spoken via telephone with Jim McManaman, the overman at the 7,800-foot level. McManaman had assured him the tracks down to that transfer point were clear and the air to that depth was breathable. Beyond that, there was no way of knowing what conditions were or how dangerous they might be. The rescuers got the first tentative answers to both questions when they encountered a group of dazed miners who were slowly making their way up to the surface. These men advised Calder that the concentrations of firedamp were hazardous just a short distance into the 13,000- and 13,400-foot levels of the mine. The Bump had caused extensive damage, and conditions below were perilous. This news wasn't at all reassuring. But Calder and his crew pressed on. After reaching the transfer point at 7,800 feet deep, they walked through the tunnel that led to the back slope. There they began their descent into the areas of the mine where they knew they'd find many deceased miners and where the Bump's devastation was sure to be widespread.

At 11:30 p.m.—two and a half hours after they'd left the surface—after numerous stops and starts, Calder and the rescue crew finally reached the entrance to the 13,000-foot level. It was here that the full extent of the damage caused by the Bump came into focus. The rescuers encountered the overman and a group of eleven miners, many of whom were in rough shape. Several had suffered injuries when huge sections of the mine floor heaved, felling the men or flinging them against the mine roof. Mind you, that wasn't the only trauma they'd suffered. Virtually all the survivors had also blacked out for various periods after having breathed firedamp. The Bump had loosed huge volumes of coal dust and toxic gases; fortunately, the ventilation system was still pushing fresh air into many areas of the mine, and as yet, there was no carbon monoxide in the mix.

The overman knew this, and he was understandably eager to get to the surface. But before departing, he shared with George Calder what he knew about conditions along the 13,000-foot level and on the two levels below—survivors he'd talked to had told him there were many dead men down there, and parts of the levels at 13,400 and 13,800 feet were now impassable. That information, limited though it was, sent a chill down Calder's spine. He now had a better idea of the difficult conditions the rescuers faced.

With firedamp levels spiking dangerously, the members of the rescue crew were forced to scuttle around on all fours. While doing so, they discovered it was impossible to access the coal face on the 13,000-foot level. That prompted the rescuers to split up and fan out. Some of the men descended to 13,400 feet while others continued all the way down to 13,800. Surprisingly, they found that the air on both these levels was breathable, for the time being at least. That was good news. Beyond that, the news was mostly bad. Large areas of the mine were impassable. "The floor just came up and smashed into the roof," said Calder to one of his crew. "How could *anyone* survive this?"[3]

That was an excellent question, and it was one all the rescuers were thinking. What's more, Calder's worst fears soon materialized, for he and his crew began coming across the bodies of men who'd perished in the Bump. Sadly, recovering their corpses would have to wait. For now, the rescuers' priority was locating survivors, and before long they found some. About 300 feet up the coal face on the 13,400-foot level, the rescue team came upon a solitary miner who was partially buried under a coal fall. After freeing the man and sending him to the surface, the rescuers continued exploring the coal face. "It was one o'clock [in the morning] when we reached the top of the 13,800-foot wall where other men were freed to go to the surface," George Calder would recall.[4]

NOT LONG AFTER the barefaced rescuers entered the mine, the first two of the five-member teams of draegermen who'd eventually be involved in rescue efforts were suited up and ready to go below. They did so shortly after nine o'clock.[5]

Calder had sent word to the surface that there were many injured men who couldn't be moved until they received medical aid to stabilize their conditions. As a result, Dr. Arnold Burden, who'd been pacing the floor of the DOSCO offices while he waited for permission to enter the mine, quickly got ready to accompany one of the teams of draegermen. Remember, Burden had worked in the DOSCO mine for two summers when he was a Dalhousie medical student. He'd also gone down into the mine with rescue crews in the wake of the 1956 explosion that devastated the No. 4 mine. Despite having been overcome by firedamp at that time, Burden remained ready to do anything and everything he could to save lives. "I was outfitted with hat, boots, coveralls, lamp, and belt and went down . . . along with two Draeger teams," he would later write. "Since I didn't work for the coal company, an official was sent with me—Dan O'Rourke. So, Dan, myself, and the trip driver were three barefaced men along with the ten Draegermen who were equipped with breathing apparatus."[6]

Their group's descent of the main slope went smoothly, and initially it seemed routine. Disarmingly so. It wasn't until Dr. Burden and the draegermen were traversing the tunnel to the back slope that they got their first real inkling of how bad the situation was below. They ran into a group of bedraggled, black-faced miners who were shuffling along like zombies while painstakingly making their way to the surface. Some were limping; others were bleeding from cuts. Still others had suffered head traumas or injuries to their arms and backs. Among the walking wounded was one of Burden's relatives. Peter Amon was the brother of Burden's stepmother.

Grim-faced draegermen, with their gear on, head into the No. 2 mine. (Clara Thomas Archives and Special Collections, York University)

"Does anyone need emergency medical treatment?" Burden called out. When no one replied, the doctor looked directly at Amon and asked him if he was all right.

"[Peter] nodded and kept going. After a few steps, he turned around and warned, 'It's very bad down there,'" Burden would remember. "That was all I needed to hear. When a miner says it's bad, look out."[7]

All too soon, it would become frighteningly clear just how bad it really was. Burden and the draegermen hadn't gone very far down the back slope before the flames in their safety lamps grew dim and then died. That set Burden sweating, and suddenly he felt numb. He knew that when there wasn't enough oxygen in the air to sustain fire, the firedamp level was dangerously high. In that instant, Burden felt very vulnerable, for rather than turning back, the draegermen pressed ahead. The mine's ventilation system was still working, and knowing that firedamp is lighter than oxygen and rises, the draegermen reasoned that the air on the mine's lowest level likely would be breathable. Fortunately, they were correct. Dr. Burden and his bare-faced companions breathed a sigh of relief—quite literally—when they realized the draegermen had been right.

Once they began to explore the 13,800-foot level, Burden discovered that sections of the level were still passable. However, it was a different story when they got to the coal face. "The roof had collapsed, and all we could see was rubble, except for a crawl space big enough to get to the bottom of the 13,800 wall," said Burden. "The wall itself was crushed and the packs demolished."[8]

The first miner the doctor encountered here was beyond help. A large section of the mine floor at the middle of the coal face had heaved up and crushed the man against the roof, killing him instantly. Tommie Tabor, the baseball-loving family man, had suffered the same fate. "Anyone working that section of the wall had been instantly entombed. There was no question," said Burden. "I felt utterly helpless, knowing that their invisible bodies were only feet away."[9]

Shaken though he was, Burden pressed ahead in his rescue efforts. After crawling through a small opening that gave them access to the rest of the level, Burden and O'Rourke came upon a half-dozen frightened, traumatized miners. While several of them were in dire need of medical attention, these men working down here were the lucky ones. Most of the dead had been working on the coal face, halfway up the wall. When the Bump hit, the coal seam had shot sideways. The effect had been devastating. Everywhere he looked, Burden saw the evidence: crumpled metal pans, collapsed timber packs, men's tools, lunch boxes, and blood. There was lots of blood on the coal.

Red rivulets that trickled down the coal face served as grim reminders of where the bodies of the dead were now entombed. In order to provide markers for the recovery workers who'd come here later, Burden and O'Rourke used chalk to mark the spots with big Xs; otherwise, when the blood dried, it would become invisible against the coal. This task was emotionally draining for Burden; he realized it was almost certain he'd known many of the men who'd died here. Some of them likely would have been his workmates in those two summers he'd worked in the mines. Others among the dead doubtless were his patients or his neighbours. "My God," Burden would remember thinking, "we've got to get to someone who's alive."[10]

He did. At one point, he came upon a group of rescuers struggling to free an injured miner who was buried in coal up to his chest. Although the man's face was black, the doctor somehow recognized him. It was Leon Melanson, the brother-in-law of Percy Rector—who unbeknownst to Burden or Melanson just then was so horribly trapped with a group of men at the top end of the 13,000-foot coal face.

When Burden asked Melanson where it hurt worst, the injured man grimaced. Through clenched teeth, he replied, "Everywhere, doc!"

Retrieving an ampoule of Demerol from his medical bag, the doctor injected the opioid pain medication into Melanson's shoulder, he only suitable part of his body that was easily accessible. That done, while efforts to free the trapped man continued, Burden went off to look for any other survivors. He was still doing that a couple of hours later when word came that Melanson's rescuers wanted the doctor to return immediately to "cut off a leg." Puzzled by this request, Burden asked for more information. He was then told that the leg of a dead man buried alongside Melanson was across Melanson's chest and hooked under one of his armpits. Burden was incredulous. "Having seen people trapped in bombed out buildings overseas [during WWII], something told me that this could be Leon's leg, and not that of his buddy who was completely buried behind and underneath him," he said.[11]

After giving Melanson another shot of Demerol, Burden declined to amputate the leg. Instead, he told the rescue workers to keep digging. They were skeptical, but they trusted Dr. Burden, and so they did as he'd suggested. As it turned out, the good doctor was correct. When the rescuers finally uncovered Melanson, they found it was his own horribly broken leg that was splayed upward and across his chest. He was carted to the surface and rushed to hospital with two black eyes, a broken nose, and a leg that was broken into more pieces than a jigsaw puzzle. Melanson was destined to survive his horrendous ordeal; however, there was no saving his leg. He would lie in a hospital bed in Springhill until doctors there decided he should be transferred to Halifax, where unfortunately for him his leg would be amputated a couple of weeks later.

ARNOLD BURDEN'S EMERGENCY medical efforts didn't end with Leon Melanson. The doctor was as tireless as he was fearless. At one point, draegermen exploring the 13,000-foot level sent word they needed Burden's help. "Come quickly," they said. The draegermen had located three injured miners who were trapped in a small, gas-filled cavity. It was accessible only to rescuers who could squeeze through a small gap in the rubble, up near the roof. There was less than 18 inches of clearance. The draegermen, wearing their bulky breathing gear, couldn't hope to get through a gap that small. However, a barefaced diminutive man who could stay low enough to escape the firedamp might. Burden, who was five foot six and tipped the Toledo scales at 120 pounds, was ready to give it a try.

Dr. Burden carried his medical bag into the mine with him when he joined in efforts to rescue injured miners. (Courtesy of the Burden family)

After scarfing down a chocolate bar that one of the barefaced rescuers had given him—the first food Burden had eaten since he'd gone down in the mine—he went to work. The rescue was as difficult as it was dangerous. Gas drove Burden and a companion back three times. It was only on their fourth attempt that they managed to scramble up and over a crumpled coal pan and were able to reach the trapped miners. Once they'd done so, they succeeded in pulling out two of the injured men; unfortunately, there was a problem with the third. He was "out of it" and was thrashing around so violently that Burden had no choice but to inject him with a shot of tranquilizer. "He relaxed all right, and so did his bowels," Burden reported. "The poor fellows pulling on [the man's] legs said it was worse than mine gas."[12]

That malodorous incident was the end of Dr. Burden's initial efforts to rescue Bump survivors. By the time he returned to the surface, fourteen gruelling hours had passed, and he was physically exhausted and emotionally drained. After getting cleaned up at the wash house, Burden drove home intent on having something to eat and then sleeping. However, after stumbling into the house, taking a bath, donning pajamas, and grabbing a bite to eat, he couldn't sleep. He found it impossible to forget the nightmarish incidents he'd witnessed and been involved in. When Burden finally drifted off, his sleep was interrupted after just two hours. A newspaper reporter called with questions for him. Unable to go back to sleep after the interruption, Burden got dressed and returned to the mine. It was all for naught. DOSCO officials told him there was nothing more he could do just then. Rescue operations had ended. All the unexplored areas of the mine were still inaccessible because of the heaved floors, collapsed roofs, and falls. The only rescuers who were still below were tunnelling to reach the places in the mine where missing men were thought to be, or they were getting on with the grim work of recovering bodies.

Remarkably, by five o'clock on Friday morning, 81 of the 174 men who'd been working the afternoon shift at the mine when the disaster happened had made it back to the surface. Those miners who'd escaped serious injuries but were in need of medical care were receiving it at the town armouries, which had been turned into an ersatz care centre and morgue. Nineteen men who'd suffered broken bones or serious internal injuries were patients at All Saints' Cottage Hospital.

"I didn't see my father at all. He was there, operating on miners who'd been rescued, and looking after the injured," recalled Anne Murray, the daughter of Marion and Carson Murray. "There were those patients Dad couldn't look after, and so he'd have them helicoptered out [to Halifax]. Dad slept at the hospital, and when he came home it was only for a few minutes at a time. I didn't see much of Mom either. She'd been a nurse, and she dusted off her skills to work at the Armouries, where they were looking after injured miners."[13]

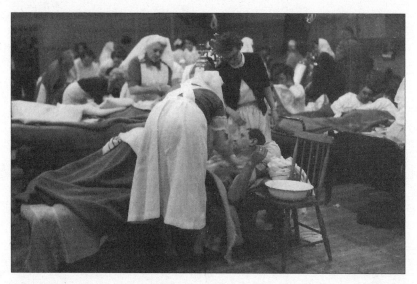

The Springhill Armouries served as a makeshift care facility when the number of injured miners overflowed All Saints' Cottage Hospital. (Nova Scotia Archives)

CHAPTER 16

"NO DUNGEON SO DARK AND SO DIM"

Friday, October 24, 1958
5:00 A.M.

A COLD DRIZZLE WAS FALLING AS DAWN NEARED ON THE morning after the Bump. The ground around the pithead of the No. 2 mine, along the path to the lamp cabin and wash house, and in the nearby parking lot had become a quagmire. With the unseasonable warmth of the previous day now long gone, and the sun shrouded by a thick veil of clouds, autumn had returned to Springhill.

Meanwhile, the scene at the mine had taken on the look and feel of an event. Fire department and civil defence officials had set up loudspeakers and floodlights. Police had strung a cordon of ropes around the hoist shed to keep away townspeople and ill-mannered newspaper photographers. (The use of flashbulbs was strictly forbidden at the pithead because of the potential for explosions.) During the night, cars with their headlights on had parked on the nearby baseball field, lighting the outfield area so a navy helicopter could land and deliver a load of plasma and other Red Cross medical supplies. Meanwhile, a fleet of ambulances that had arrived from Amherst, Moncton, and Halifax was gathered in one corner of the mine's parking lot, while in another the Salvation Army and Red Cross had erected large tents where emergency workers—twenty-one of them who'd driven from Saint John, New Brunswick—were serving coffee and food to exhausted rescue workers, locals, and even the out-of-town print journalists and photographers, whose intrusiveness was already starting to rankle the locals.

Among the first emergency workers to appear at the mine site were "the familiar berets of the Canadian Legion Ladies' Auxiliary, the blue serge and red trim of the Salvation Army uniforms, and the white armbands bearing the emblem of the Canadian Red Cross."[1] The women were working on their own initiative. The chairman of the Springhill branch of the Red Cross, Edwin McKinnon, was among the missing miners. Sadly, his body would be recovered from the mine a few days later.

Despite the cool, damp weather, the crowd that had rushed to the pithead in the minutes after the Bump maintained their mostly silent vigil; those who remained on watch were still numbed by the shock of what had happened. "As the minutes of waiting lengthened . . . there was no definite announcement made to the crowd regarding the safety or otherwise of the men below,"[2] as the *Springhill-Parrsboro Record* reported. Anytime a barefaced miner or one of the draegermen emerged from the hoist shed, the crowd surged forward, and the women and girls in the group called out questions about specific men. At this point, these were questions for which there were no answers.

Over at the wash house—a soggy football-field-distance to the west—a *Boston Globe* reporter was intent on quizzing a couple of exhausted draegermen. Hoping to catch some badly needed shut-eye, the two men had flopped down on the cots that were set up there. "You know, there just can't be any more [men] alive in the mine," one of the rescue workers told the nosy journalist. "Those guys who got out at the beginning . . . they were sure lucky. Next time we go down, it'll be just for bodies."[3]

That same grim view of the situation was repeated a few minutes later by Bill Totten, one of the miners who'd escaped in the first few hours after the Bump. The twenty-seven-year veteran of the mine addressed a throng of reporters who were gathered in another room in the wash house. Totten had been working at the 13,400-foot level when the earth shook. Although the upheaval had left him badly shaken—physically and mentally—he'd gone back down into the mine as a member of one of the barefaced crews. Then, having done all he could to rescue co-workers who were trapped, Totten was eager to talk about his experiences and offer his assessment of the conditions underground. "You can't imagine what it's like," he said. "Crawling through a little hole just big enough to pass through. Every time there's a rumble or a stone falls, you get ready for another bump. That would finish us. The heat and the smell . . . digging out those bodies. It's awful."[4]

Another miner who'd also escaped the mine soon after the Bump and then joined in rescue efforts echoed Totten's words. "There's no room down there. I have cuts all up my arm from scraping it on the rock," Stan Pashkoski said. "We crawl along in a line, about a dozen of us. It takes a long time, sometimes a whole shift of eight hours, to go a couple of feet."[5]

Despite this, Pashkoski remained confident that many of the missing men would be found alive. That was, *if only* rescuers could reach them in time. If they didn't, he vowed it wouldn't be for lack of effort. Springhill miners didn't give up when they were trying to rescue one of their own.

While everyone wanted to share in Pashkoski's optimism and to believe those who didn't give up couldn't be beaten, keeping the faith wasn't easy. Everyone knew there would be difficult days ahead. The first reminder of that grim reality came when the first body that rescue workers had recovered was removed from the mine. That happened in the pre-dawn light of this sodden Friday morning. A misty rain was falling at about 5:40 a.m. when a draegerman standing in the door of the hoist shed signalled to the police and army. He wanted them to keep the crowd from surging forward as an ambulance pulled up. Onlookers watched in dazed silence as a couple of draegermen carried a stretcher out to the vehicle's open back door. On that stretcher was a body draped with a white sheet.

"It's Harry Halliday!" someone whispered. When someone else repeated that news, it instantly became a buzz that swept through the crowd. And so it was that the world learned fifty-three-year-old James Harry Halliday was the first victim of the disaster to be recovered and identified.

Halliday had been among the men working on the coal face at the 13,800-foot level when the Bump hit. He and Fred Hahnen, a thirty-eight-year veteran of the mines, were about six feet apart when it happened. "We were kidding about who was getting the most work done. You know, just one of those things to pass the time," Hahnen recalled in a hospital-bed interview with a newspaper reporter.[6]

The Bump had sent both Halliday and Hahnen head over heels and left them buried under an avalanche of loose coal and rock. How long Hahnen was unconscious, he couldn't say. But he was fortunate. Co-workers found him and dug him out in time to save his life. Halliday wasn't as fortunate. The cause of his death, as determined by Dr. Burden, was MSCC—"multiple severe contusions and crushing." That abbreviation tragically would become all too familiar to Springhill residents in the coming days.

It happened that Harry Halliday's wife was in the crowd at the pithead when his body was removed from the mine and transported to the Springhill Armouries, which was being used as a temporary morgue. When she heard the identity of the man whose corpse was being loaded into the ambulance, Eva Halliday began weeping uncontrollably. Friends did their best to comfort her and then escorted her to the Halliday family home on Main Street. There waiting in the living room were the Hallidays' two daughters and Harry's mother, who was seventy-four years old and blind. The grey-haired grandmother had heard via the town grapevine that a man's body had been recovered at the mine; somehow, she sensed who it was. She'd always feared this day would come—the day her son would die in a mine accident. The elderly Mrs. Halliday sat in numbed silence, tears trickling down her cheeks, as her daughter-in-law quietly announced that Harry would be buried two days hence, on Sunday.

EVENTS IN SPRINGHILL were now happening at a dizzying pace. Top DOSCO officials, including Sir Roy Dobson, the head of DOSCO's corporate parent, A.V. Roe UK, and Crawford Gordon, the president of its Canadian subsidiary, A.V. Roe Canada, were en route to Springhill from Toronto; when the Bump happened, they'd been there attending the company's annual corporate meetings. At the same time, Nova Scotia premier Robert Stanfield and several members of his cabinet were coming from Halifax. (Politicians then—just as is still very much the case today—never miss a photo op.) Meanwhile, mere minutes after the recovery of Harry Halliday's body, a car driven by Harold Gordon, DOSCO's vice-president and general manager, pulled into the mine's muddy parking lot. Gordon had made the 300-mile nighttime drive from Sydney in a mad dash; having been alerted to the emergency, police gave him a free pass to speed.

At six feet two inches tall with chiselled features, a dimpled chin, and a thatch of swept-back thinning hair, Gordon was an imposing physical presence. His looks matched his no-nonsense, hands-on approach to life. He took a personal interest in DOSCO's mining operations and in the welfare of the miners who worked for the company, "his boys."

Gordon, who was fifty-nine, had been in and around mines all his life. Scottish born, he was seven years old in 1907 when his family had immigrated to Canada so his father, an executive with DOSCO, could take up a management position in Sydney. In his youth, Gordon had worked as a pit boy in a coal mine before he attended McGill University. After he earned a degree in mining engineering in 1923, his father helped him land a job with DOSCO back in Cape Breton. Gordon then set about working his way up through the management ranks. The thirty-five years he'd spent behind a desk hadn't dimmed his ardour for underground work. In 1956, Gordon led the rescue teams that saved dozens of miners after the devastating explosion that rocked the No. 4 Springhill mine. Now he was ready to don draegerman gear once again and do whatever he could to direct rescue operations and save any men trapped in the shattered No. 2 mine. As it turned out, there wasn't much he could do this time around, at least not initially.

Gordon returned to the surface at 10:30 a.m. after having spent more than three hours underground. He'd explored parts of both the 13,400- and 13,800-foot levels; however, he'd been unable to enter the tunnel at the 13,000-foot level, which was blocked with rubble and full of gas. This was especially frustrating since Gordon and many miners in the rescue crew thought most of the missing men who might still be alive would be trapped there.

When Gordon emerged from the pithead, he was stone-faced and blackened with coal dust, his jaw clenched. He strode along the muddy path that led to the wash house. Gordon said nothing, but the fact his customary bounce was gone from his gait told onlookers he was downcast. Why that was so would soon become all too clear.

When Gordon found C. Arnold Patterson, DOSCO's Montreal-based public relations director, waiting for him in the wash house, he instructed Patterson to arrange a media conference for one hour hence. Gordon had bad news to share with the 137 journalists—Arnie Patterson kept a count—who'd descended upon Springhill and were ravenous for any information that would provide fodder for the breathless dispatches they were churning out. News of the Springhill mine disaster was generating headlines and attracting attention worldwide.

In addition to the expected rush of reporters that Springhillers were accustomed to seeing whenever there was trouble at the mine, a television crew appeared in town. The live, on-the-spot news reporting they did was a first for Springhill and for Canada. There'd been a radio news reporter on hand in April 1936 when a cave-in trapped three miners in a gold mine at Moose River, Nova Scotia (an hour's drive northeast of Halifax). Days passed as rescuers struggled to find a way into the mine that wouldn't cause a disastrous cave-in.[7]

On April 20, eight days after the start of the crisis, a mustachioed, pipe-smoking reporter with the Canadian Radio Broadcasting Commission (the forerunner of the CBC) named J. Frank Willis arrived on the scene. Making use of the shared telephone "party line" that was the only one available in the area, Willis forged international broadcasting history. For almost three days, each half-hour, he aired live two-minute updates on the rescue efforts. In doing so, Willis created North America's first twenty-four-hour "news event." It's guesstimated that as many as 100 million listeners across North America and around the world heard his dramatic live reports.[8] Radio news reporting would never be the same. A similar phenomenon happened when that CBC television crew came to Springhill to report on the town's latest mining disaster.

The CBC had launched its television broadcasting service only six years earlier. On September 6, 1952, Montreal station CBFT went on the air, and two days later Toronto station CBLT followed suit. Two years later, CBC television programming came to Nova Scotia when, in October 1954, station CJCB began operations, and then on December 20, 1954, CBC station CBHT began broadcasting over Channel 3 in Halifax.

Until 1958, CBC had a monopoly on domestic television broadcasting in this country since federal law prohibited the creation of private television networks. Television broadcasting was still very much in its infancy in 1958, news broadcasting even more so. On July 1 of that year, Dominion Day, a network of 139 microwave towers—the world's longest—began beaming television signals across the country to forty privately owned stations and eight that were operated by the CBC—those two stations in Nova Scotia being among them.[9]

It was impossible for anyone to miss the bulky Canadian Broadcasting Corporation cameras, the technicians, the glaring floodlights, and the trench-coated reporters with microphones in hand who showed up in Springhill in the wake of the Bump and eagerly began interviewing mine officials and locals. First on the scene was Keith Barry from CBHT, and then as interest in the story grew, Lloyd MacInnis and Jack MacAndrew flew in from Toronto. "This was the first time that television covered such an event," MacAndrew would recall many years later.[10] He was right about that.

Before this time, television cameras had never ventured outside the controlled studio conditions to broadcast the drama of a live news event. The technical quality of the CBC broadcast was poor, yet despite the darkness of the video and the garbled audio that sounded as if it were coming to earth from another planet, it made for exciting viewing.

Today, we tend to take for granted the technology that makes such live news reports possible—whether broadcast over the airwaves or streamed online; at times, the report even seems quaint, maybe a tad primitive. What's surprising nowadays is that with smart phones being ubiquitous, we're taken aback if a video of every event or disaster isn't readily available to view on social media or television. The world was a very different place in 1958. A live television broadcast of on-the-spot news "from the field" was cutting-edge. And it was very much a novelty.

CBC newsman Jack MacAndrew (*centre*) and two of the technical crew taking a break from their remote live news broadcasting of the Springhill mine disaster. (Springhill Miners' Museum)

The live coverage of the coronation of Queen Elizabeth II on June 2, 1953, is widely regarded as the event that made television a mainstream news medium. More than twenty million people tuned in to watch the broadcast. The CBC's television coverage of the Springhill disaster didn't draw nearly that many viewers, but it is similarly memorable since it was the first news broadcast with real-time on-the-spot coverage of events as they happened. While the technical challenges were immense, the CBC production people pulled it off, going live across Canada and even beyond.

It's no exaggeration to say the public broadcaster's coverage of the Springhill mine disaster marked a milestone in television coverage of breaking news, even if those who were involved in the broadcast—like media observers generally—seem not to have been aware of it at the time.[11] That the CBC made such an effort to air its coverage of the events in Springhill speaks volumes about the perceived level of public interest. Indeed, the world was paying attention. Television networks in the United States and Europe carried reports of the disaster. The impact of this was coverage as remarkable as it was unprecedented. Television viewers globally tuned in, and what's more, they sent more than forty thousand letters and postcards,[12] as well as gifts of food, clothing, and even cash to the people of Springhill. By the end of October, almost a million dollars had been received and added to the miners' relief fund that had been established.[13]

From London, the Queen had sent a message of support to the people of Springhill and offered her condolences to those who'd lost loved ones. Prime Minister John Diefenbaker in Ottawa dispatched a similar message, as had Nova Scotia premier Robert Stanfield before he left Halifax and rushed to Springhill. While such sentiments were appreciated, they did little to brighten the mood in the town. It seemed the only news coming out of the No. 2 mine was bad news.

When an overflow crowd of journalists packed a meeting room in the DOSCO offices for Gordon's press conference, they listened in stunned silence as the manager gave his assessment of the situation in the No. 2 mine. Speaking slowly and in a voice that at times was quavering and barely audible, he delivered a frank assessment of the situation underground. "I regret to tell you that I consider there's no hope for any of the men on the 13,400- and 13,800-foot levels," he said.[14] After pausing a moment, he then added a chilling afterthought: he had only "the faintest of hopes" for the eighty-three men who were still missing.

Louis Frost (*middle*), DOSCO's chief mining engineer, and company vice-president and general manager Harold Gordon (*to Frost's left*) held daily press conferences at the DOSCO offices as company public relations chief C. Arnold Patterson (*in hat*) looked on. (Nova Scotia Archives)

None of this was what anyone wanted or had expected to hear, certainly not the miners or the people of Springhill. Most locals didn't believe Harold Gordon's words. They didn't want to, and they refused to do so. In houses from one end of town to the other, the families of the missing miners kept the lights on, figuratively and literally. They prayed their loved ones were still alive and would be coming home. Even when she went to bed at night, Margie Kempt kept all the lights burning in the Kempt family home on Herrett Road. Her husband, Gorley, the father of her two children, was among the missing, and never—not even for one moment—would she let herself think he wasn't alive and wouldn't be coming home. "Mum was emphatic that despite all the bad news, Dad would survive, and so she was symbolically keeping the lights on day and night," Billy Kempt recalled. "She said that if she turned them off, it would look like she didn't think Dad was coming home."[15]

Hope. That was the vital word. It was all that Margie Kempt and the other wives and families of eighty-three missing men had to cling to. They couldn't—no, they *wouldn't*—ever give that up.

CHAPTER 17

OUT OF EVERYTHING BUT HOPE

FRIDAY, OCTOBER 24, 1958
5:30 A.M.

WHAT DO PEOPLE THINK ABOUT WHEN THEY'RE FACE TO face with their own mortality? In the romanticized Hollywood version of what happens, the dying person utters something pithy and memorable—"famous last words"—before shuffling off this mortal coil, as William Shakespeare put it. However, for the most part, the reality of death is a lot more banal than that.

The researchers who study such things tell us that when most people are drawing what they believe are their final breaths, they take stock of what they've done or haven't done in this life. Last thoughts turn to home and loved ones. That was certainly the case for the ill-fated coal miners who lay dying in a firedamp-filled pit in Fraterville, Tennessee, in 1902. One poor soul left a note advising his sons, "Never work in coal mines." Another miner scribbled a farewell note to his wife in which he said, "If I don't see you any more [sic], bury me in the clothing I have. I want you to meet me in heaven. Goodbye. Do as you wish."[1]

None of the dozen men, the Twelve, who were trapped at the bottom end of the 13,000-foot coal face, at the 13,400-foot level of the No. 2 Springhill mine, were ready to write a farewell note. Not yet, anyway. However, nine hours after the catastrophic bump that shattered the mine and upended life, their mood was as glum as mud. The gravity of their predicament had started to weigh heavily on the men's minds. Understandably so. After all, here they were in the dark, entombed three-quarters of a mile underground. They had very little food or water. No one on the surface knew they were alive; and even if anyone did, no one knew exactly where to find them. Making matters worse, the air the men were breathing continued to be fouled periodically with toxic gas and with the constant, growing stench of the corpses that were already starting to putrefy in the warmth of the mine. The odour was overpowering. If there is such a place as living hell, this was it.

Harold Brine contemplated the misery of his situation as he lay on the uneven, sloping surface of the mine floor. It was a hard, uncomfortable bed, and sleep hadn't come easily for him during the night or at any time. Brine had retrieved a watch with a luminous dial from the wrist of a corpse that was lying in the rubble, about 30 feet up the coal face; Brine knew the dead man and vowed that if he made it out of there alive, he'd deliver the watch to his widow. But for now, that precious timepiece was Brine's only tangible link to the world of the living. Tellingly, when the time fell back to Daylight Standard Time on the night of October 26, the ever-conscientious Brine would remember to turn his watch back an hour—400 feet away, so too did Maurice Ruddick, who also had a watch with a luminous dial.

"Imagination," the French writer Joseph Joubert once said, "is the eye of the soul." Lying there in the dark in the wee hours of this Friday morning, Harold Brine allowed his mind to fly drone-like a couple of miles to the west of Springhill. In his mind's eye, he could see the bungalow he was building. That house and his family were at the essence of his life, his soul. This early in the morning, he imagined that his wife Joan was still asleep with their infant daughter, Bonnie, snuggled in next to her. Brine wished with all his heart he could be there at home and in bed with them. In the next moment, the realization hit him: he might never go home again. He might die here in the mine.

If that happened, he felt certain that Joan would be all right. She was bright and pretty, and at twenty-four years old, she still had plenty of time to remarry or find a job, he reckoned. Maybe both. Then too, Joan would also collect widow benefit cheques— fifty dollars a month for life or until she remarried—and she'd also receive twenty dollars per month for their two-year-old daughter. "The one thing I really regretted when I thought about it was that if I didn't get out of there, Bonnie was too young to remember me," said Brine.[2]

That possibility was painful even to contemplate. Doing so stiffened Brine's resolve to live. Dammit, he decided, he was going to survive this ordeal. Somehow. But he knew it wouldn't be easy. As Levi Milley had bemoaned, "We're out of everything but hope." And even their hope was fading fast.

WITH EACH PASSING hour, the headlamps of Harold Brine and his co-workers were growing ever dimmer. One by one, they faded to black. And as they did, it became ever more difficult for the Twelve to stay upbeat. Some of the men were already on the verge of giving up. Brine could hear it in the rising tide of desperation and fear in their voices, especially that of Larry Leadbetter. The youngest man among them was straddling that thin line between composure and panic.

Tunnel vision and loss of perspective are panic's companions. Once it takes root—like fear, anger, or suspicion—panic becomes all consuming and spreads with the speed of summer lightning, especially in a confined space when nerves are frayed. That's why it didn't help when in a moment of despair Ted Michniak, who was in a bad way with his dislocated shoulder, told Levi Milley, "We're not going to make it out of here." It was as though he had broken a taboo.

A defeatist message wasn't something the other men wanted to hear, even if they themselves were starting to think it. Joe McDonald, ever feisty despite his injuries and his constant pain, was the first to speak up. He'd done some boxing in his youth, had a quick temper, and liked to think he still threw a solid punch. He was in no condition to fight, of course, and Ted Michniak was a good friend and was lying not far away, but that didn't stop McDonald from barking at him.

"Damn it, Teddy! I'm not ready to throw in the towel, nor is anybody else here," said McDonald. "I don't care if you think you're not going to make it out of here, but the rest of us will. So stop with your negative talk."

Fortunately, when Michniak didn't respond, the tension that filled the chamber was short-lived; after a few fraught moments, McDonald took a deep breath and broke the bristly silence. "Let's have something to eat," he suggested.

Out of the darkness came the voice of Bowman Maddison. "Anybody know what time it is?" he asked. "I like a late breakfast."

That comment prompted Caleb Rushton to check his watch. He did so and announced that it was almost nine o'clock, eliciting another lighthearted quip from Bowman Maddison. "I don't get up this early," he chirped.

No matter. The men had decided it was time for "an early breakfast." Not that there was much to eat. Their meagre store of food consisted of a couple of stale sandwiches that the men had scrounged up while rummaging through the lunch boxes of dead co-workers. What little water they had, a pint or two, wasn't enough to fill a water can. The men had entrusted their supplies to Levi Milley. He was one of the elders in the crew.

Milley had been a miner for two-thirds of his forty-seven years. As a young man, he'd dreamed of being a travelling salesman or a chicken farmer. Neither career option had worked out for him, although he still raised chickens and cackled with delight when he talked about his flock of fifty-five birds. He'd instead gone to work in the mine to help his widowed mother feed the family. Milley, a lean man with grey, thinning hair and a quiet demeanour, had a certain presence, and he was respected. Co-workers regarded him as being "a straight shooter," someone who was trustworthy. As a result, it was to him that they entrusted their tiny store of food and water. They believed Milley would safeguard and apportion it fairly. He was diligent in doing both. This morning, he carefully broke one of the sandwiches into a dozen bite-sized portions. Each man got his tiny share along with a sip of water from an empty Aspirin bottle that Ed Lowther provided. It wasn't much, but any food or water was better than none, and having it raised the men's spirits.

Afterward, when it was quiet again, Fred Hunter slipped away into the darkness to visit the body he wrongly believed was that of his brother Frank. Doing so was an ordeal. Hunter's injured leg had ballooned to twice its normal size, and the pain he felt was excruciating. Adding to his misery was the stench of the decaying corpse he visited. In the heat of the mine, the smell was already becoming overwhelming, so much so that Hunter tried to lessen its unpleasantness by breathing through his mouth rather than his nose. That didn't work, and so he beat a hasty retreat.

At the same time Fred Hunter was scrambling to rejoin the others, Bowman Maddison, Levi Milley, and Hughie Guthro were also up and around, moving along the coal face. The trio turned on their lights and went exploring yet again. They did so in hopes of finding more food and water, some lamp batteries, or maybe even an opening that would lead to a way out of there. They had no luck in any of those regards. All they found was the reeking body of yet another dead miner. They left it where it was and quickly returned to the cavern where their companions sat waiting in misery.

The rest of the day passed quietly and uneventfully. Then late that afternoon, Bowman Maddison broke the silence. Unexpectedly, he called out, "What's that noise?"

In that moment, eleven other men who'd been silently lying in the dark, languishing in that shadowy netherworld between consciousness and dreams, were instantly awake. They listened intently until they heard it too. From somewhere far away came a faint *rat-a-tat-tat*. It sounded like the noise made by a chipper, the air-powered chisel that miners used for tunnelling. "Do you hear that? They're coming for us!" shouted Maddison. He was gleeful.

His words were like a group adrenalin shot. Suddenly, the men had renewed hope. They strained to listen and make sense of what they were hearing. However, the noise lasted only a few more seconds before it died away. When it did, the disappointment was palpable. The silence that followed was deeper and emptier than ever. "They've quit for the weekend," somebody said. "They probably think we're all dead."

That crushing thought lingered in the mind of each of the Twelve. They passed the next couple of hours in forlorn solitude. Finally, around six o'clock, that time in the afternoon shift when the miners usually opened their lunch boxes and began their dinner break, Gorley Kempt stirred. He went off in the darkness on his own to forage for food or water. His luck was good this time out. When he returned, it was with a lunch box he'd found. In it were a few stale bread crusts that were gritty but edible. Skip the dishes. This was dinner for the Twelve tonight.

A crowd of Springhillers—many of them the wives and children of missing miners—rushed to the pithead the night of the Bump. Some of them remained there until recovery workers retrieved the last body from the mine. (Clara Thomas Archives and Special Collections, York University)

CHAPTER 18

"IT'S FOR THE LORD TO DECIDE"

———

FRIDAY, OCTOBER 24, 1958
6:00 A.M.

BY EARLY FRIDAY MORNING, TWO THINGS HAD BECOME abundantly clear to all the men who were trapped in the devastated No. 2 mine. One was that the damage wrought by the upheaval was even more extensive and deadly than anything they'd initially feared. The second was that the odds were slim that rescuers would reach them quickly. Time was a relentless, unforgiving enemy. The page one headline in the day's "extra edition" of the *Halifax Chronicle-Herald* proclaimed, "All Efforts to Reach Men Foiled by Gas."

More than half a mile deep, for better or worse, the trapped miners were unaware of any of this. There in the darkness of their subterranean isolation, they were preoccupied with even more immediate life-and-death concerns. Up near the top end of the coal face on the 13,000-foot level, Maurice Ruddick was struggling to make a difficult decision that, once made, couldn't be walked back. Having taken it upon himself to do all he could to look after Percy Rector, Ruddick had decided to give him the last of the

Aspirins that the Seven had found in their pockets.[1] After placing four pain-relief tablets between Rector's dry, parched lips, Ruddick provided him with a tiny bite of a sandwich and a sip of water. None of this made much difference to Rector, of course. It was impossible to relieve his suffering, not with his crushed right arm still hopelessly pinned between the timbers of that collapsed pack. Rector was hanging there, unable even to sit or find any measure of relief from his pain. What was even worse, having no other way to relieve himself, he'd urinated and soiled himself. The indignity of it all, not to mention the smell, was horrible for one and all. The only time Rector's heart-rending moaning stopped was when he passed out because of exhaustion or as a result of having inhaled too much of the firedamp that periodically came and went, fouling the air in the mine; unlike the other men, he couldn't drop down to the floor to avoid the gas.

Mercifully, all was quiet for a few minutes after Percy Rector swallowed the last of the Aspirins. For a few minutes, at least. It wasn't long before Rector began sobbing again, this time with renewed intensity. In the confined space in which they found themselves, the other men had nowhere to escape, no way to tune out or ignore his heartbreaking cries. "I can't stand it anymore. For God's sake, cut off my arm. *Please!*" Rector wailed before he passed out again.

In the silence that now enveloped the chamber, the discussion about what, if anything, the men could do to relieve Rector's suffering soon resumed. The reality was that none of the Seven could take much more of the poor man's wailing, yet they felt helpless; there seemed to be nothing they could do. They were paralyzed with indecision.

"What does everybody think?" blurted Herb Pepperdine.

"We've got to talk about it," said Maurice Ruddick, and so they did. They huddled together in the darkness and spoke in hushed voices for fear that Rector might overhear their conversation. There was little chance of that. He was unconscious, and when he wasn't, he was mostly delirious.

Garnet Clarke was the first to speak up. He feared if they sawed off Rector's arm, the trauma would kill him. Herb Pepperdine agreed; however, he added that there had to be something they could do. What if they performed the amputation and then shredded their shirts and used strips of the cloth to make tourniquets to bandage the stump? Maybe they would be able to stop the bleeding.

Ruddick conceded that idea might work, but he was still opposed to an amputation. "What if we do it, Percy dies, and then we get rescued an hour later? How will we feel then?"

The persuasiveness of that argument prompted the nodding of heads around the circle, and then there was more silence. It was Garnet Clarke who spoke next. Why not settle things with a vote? Maurice Ruddick didn't much care for that idea. Whether Percy Rector lived or died, he insisted, wasn't their decision to make. "It's for the Lord to decide," he said.

When no one disputed that assertion, the pressing question they were wrestling with—like Percy Rector—was left hanging in the air: what would they do next? Would they leave Percy to suffer, or would they roll the dice, cut off his arm, and pray the shock wouldn't kill him?

"We've got to make a decision. I think we should vote on what to do," Clarke said.

That seemed like the only reasonable course of action, and so the men opted to hold a straw poll. When they did, it may have been Rector's renewed crying that influenced the outcome. Or perhaps the deciding factor was Maurice Ruddick's sobering reminder that if they did opt to go ahead with the amputation, one of them would need to wield the saw.

When the decision was made, it was unequivocal. The Seven voted unanimously not to cut off Rector's arm. Better, they reasoned, to wait for the rescuers and pray they would arrive sooner rather than later.

In the wake of that vote, the men once again fell silent. Even poor Percy Rector. It was while they were sitting there in the darkness, mulling over the implications of their decision, that they heard scratching noises. Rats. That's what it sounded like; rats were a constant in the mine. The pesky vermin usually came around in droves at dinnertime, when the miners tossed bread crusts and other refuse into the gob area. If any of the Seven had thought about it, they might well have wondered why rats were still scurrying around after such a major bump. Just as rats jump off a sinking ship, they instinctively flee a mine when they sense there's a cave-in or upheaval coming. Why the rats might be around now was a question that never occurred to any of the men. They were preoccupied just then with their need for food and water. It was more than eight hours since any of them had eaten, and the men were starting to feel hungry and thirsty.

A thorough search of the debris up and down the coal face had turned up just three stale sandwiches and about a quart of tepid water. The dearth of the necessities of life was a major concern for the men, and it was a huge potential source of conflict. In the darkness, it was as easy as it was inevitable that someone would stash food or water or maybe sneak more than his fair share of whatever was available. A few angry, accusatory words were muttered and pondered, suspicions simmered, but nothing came of it. So far, the Seven were still in relatively good condition. "No one went haywire," said Ruddick. "No one went off their rocker. We were all talking sensible, ordinary conversations."[2]

The men entrusted their tiny store of food and water to Ruddick. At age forty-six, he was one of the oldest members of the group. Although he didn't talk much, he was well respected. There was an inner strength to Maurice Ruddick that many people admired. It may have had to do with the fact he was such a strong father figure—how could he not be when he loved and provided for his wife and their dozen kids? Or it may have had to do with Ruddick's singing, which had a calming effect on the others. Or perhaps it was his willingness to care for and do whatever he could to protect Percy Rector in what can only be described as trying conditions. But whatever the reason, his co-workers believed Ruddick could be trusted to apportion the food—one sandwich each day divided seven ways—and to see to it that each man got his fair share of the precious water.

DOWN AT THE BOTTOM end of the 13,000-foot wall, Harold Brine and his eleven companions in misery faced a similar dilemma. Friday had slipped into Saturday, and they'd had almost nothing to eat or drink. Today, they shared a single grit-covered sandwich and divvied up a small amount of their remaining water, which they passed around in Ed Lowther's empty Aspirin bottle. Divided a dozen ways, neither the sandwich nor the water went far. There was no Biblical miracle to save the day.

In the face of such adversity, the spirits of the Twelve faded faster than had the batteries of their headlamps. The men had done everything they could think of to find an escape route, repeatedly exploring both ends of that area of the coal face where they were imprisoned. "We tried to get out of there, but when we dug, we realized it was solid at the bottom of the wall. The floor was right up to the roof," said Harold Brine. "When we tried to climb up the wall, it was bad there too, and there was a lot of gas. When I got gassed, the boys grabbed me by my feet and pulled me back down. Then they laid me out on the floor where the air was clear. That's where I came to."[3]

In the face of such disappointments, it had become clear to the men that they had no way out. There'd be no escape for any of them. The only thing they could do was sit, wait, and pray that rescuers would find them before it was too late, before they succumbed to hunger or thirst. It was depressing to contemplate such a slow, painful death, so Caleb Rushton took it upon himself to divert their minds.

Rushton, clean-shaven, bespectacled, and devout of faith, was usually a soft-spoken and studious man. However, when needed, his voice could be sonorous. It was now, and he put it to good use, leading the Twelve in prayer and in singing one of his favourite hymns, "The Stranger of Galilee": "In fancy I stood by the shore, one day, / Of the beautiful murmuring sea . . ." he sang.

The words Rushton uttered or sang were small succour, but they brought a measure of comfort and calm to the men. One by one, they dropped off into fitful sleep.

IN THE WAKE of the vote that nixed the idea of amputating Percy Rector's arm, there was a strained silence in the space where the Seven sat and waited. And waited. What they were waiting for, other than the appearance of rescuers, none of them could say. When they'd finally had enough sitting, Garnet Clarke and his pal Currie Smith decided to set off on yet another foraging expedition. Why not? The pair had nothing else to do, and they figured if they got lucky, they might find fresh batteries for their fading headlamps or maybe the lunch boxes or water containers of some of the dead men whose bodies were now buried in the debris that clogged the coal face. Locating bodies wasn't all that difficult; the dead were beginning to decompose rapidly now, and the stench of decaying flesh pervaded the mine. Locating any lunch boxes, water bottles, or lamp batteries wasn't as easy. The headlamps Clarke and Smith carried were growing ever dimmer, and once the batteries died, the men would be left to grope around in the dark.

Clarke and Smith were crawling around on their bellies, poking into crevices and wiggling loose rocks. It was while doing so that they came upon what looked to be an inviting opening in the debris field that might be an escape route out of this area of the shattered mine. It was about 75 feet from where the rest of the men were sitting. With Smith's help, Clarke cleared an access large enough for him to crawl through. When he did so, there was a surprise waiting for him. The pale beam cast by his headlamp revealed the body of a man who was wedged into the crevice.

Clarke brushed away the dirt and loose coal that were partially covering the man's face. "My God. It's Barney Martin," he said.

At that moment, Clarke got another surprise. Martin, who was still clinging to life, stirred. The sound of someone speaking his name summoned him from his semi-comatose state. "Hello," he croaked in a voice that was barely above a whisper.

Springhill-born Byron Martin had worked in the mine for eighteen of his forty-two years of life. "Barney," as most of his mates called him, had been severely injured in a 1952 rock fall that left him with a fractured back and one of his ears almost torn off; his ear was saved by the surgical skills of Dr. Carson Murray, who'd somehow managed to sew it back on. As bad as those 1952 injuries were, they paled in comparison to the harm this new bump had inflicted on Martin. He'd been working at the top end of the coal face on the 13,000-foot level when the upheaval sent him tumbling head over heels down the longwall. The trauma had dislodged his helmet and left him wedged between two crumpled stone packs. Several large boulders and a crush of loose coal had then rained down upon his legs, pinning him and reinjuring his back. Loose coal and dirt had also buried his face.

Martin had been lucky to survive. He'd been unconscious after the Bump. When he finally awoke, he was in a world of pain. His mouth was full of dirt, and his tongue felt as dry as coal dust. What was even worse, without his headlamp, he was totally disoriented.

In the pitch black, he had no idea where he was or how much time had passed since the Bump. Adding to his panic was his inability to free or move his legs. When he tried to cry out, his voice was as weak as a newborn lamb's. The silence that enveloped him was total and terrifying. The only positive was that Martin's canteen, still about a quarter full, had somehow stayed clipped to his belt, and he could reach it. He fumbled with the lid, tore it off, and reflexively swallowed a huge mouthful of water.

Martin drifted in and out of consciousness after this. At one point, in a moment of terror and confusion, he'd gulped down the last of his water and then promptly vomited. Then the firedamp that was flooding the mine in periodic waves knocked him out. Each time he regained consciousness, the terror he felt at being buried alive filled him with fresh panic. Coal dust burned his eyes. He cried out in desperation and fear. He screamed and blindly lashed out at the rocks that imprisoned him. It was the sound of his blind, incoherent rage that the six men huddled together in the chamber with Percy Rector had heard and dismissed as rats squealing and rustling around in the gob.

Martin had struggled, clawing at the rocks until his fingertips were raw and bleeding. And then, mercifully, he had passed out yet again. He remained unconscious until Garnet Clarke and Currie Smith happened along. When he spoke, the raspy voice coming from the body of a man he'd assumed was dead startled Garnet Clarke; however, he quickly recovered. Then turning to Smith, he whispered, "Barney's not going to last much longer, but let's try to get him out of here anyway."

And so they did. It took a lot of effort for Clarke and Smith to move the rocks and other debris that had pinned Martin's legs. But once he was free, his rescuers still couldn't get him out of the crevice. He'd lapsed back into unconsciousness. The space he occupied was roughly six feet long and three feet wide. Grave-sized.

Clarke and Smith had cleared the dirt and coal away from

Martin's face so he could breathe. When he stirred, they pleaded with him to "hold on" because help was coming—even though they knew it probably wasn't. That was all the two men could do for Barney Martin. They knew he was hungry and desperately thirsty, just as they were, but they had nothing to give him. Clarke and Smith left Martin where he was when they returned to the chamber where the other miners were languishing. Even though their lights were now all but dead, the two easily found their way back to their comrades; Percy Rector was still sobbing. "We found Barney Martin," Clarke told the others. "He's in bad shape, and I don't think he's going to make it."

There was nothing more to say about that. Or anything else for that matter. The group of seven survivors had increased by one. Eight men were now almost out of food and water. Their headlamps were as dim as the embers of a dying fire. And similarly, they all knew that with each passing hour their chances of survival grew ever dimmer. There was nothing they could do other than to sit waiting in the darkness, each man alone with his thoughts. The one thought that was in every man's head as he drifted off into a fitful sleep was that if rescuers didn't soon find them, death surely would.

IT WAS SOMETIME after midnight on Saturday night when a fresh outburst of Percy Rector's cries shattered the silence at the top end of the coal face at the 13,000-foot level. Six exhausted men were suddenly wide awake again. They listened and prayed that Rector would fall silent once more so they could sleep and forget their misery. When he finally did so, the men became aware of a noise that made them all sit up and take notice. From somewhere in the distance came the sound of hammering. Six men held their breath. Sightless in the pitch black, they strained to hear every sound, for their hearts were pounding like kettledrums. This went on for several minutes. Then suddenly, the sounds of banging stopped and there was only silence again. Soul-crushing silence.

"They're coming for us," Maurice Ruddick finally said.

"Well, if they are, they're still a long way away, and they won't get here in time to save us," sighed Herb Pepperdine. "It doesn't sound like they're in much of a hurry. But why would they be? They don't need to hurry to reach dead men."

CHAPTER 19

FADING HOPES

———

SUNDAY, OCTOBER 26, 1958
12:00 P.M.

SUNDAY MORNING IN SPRINGHILL BEGAN SUNNY AND COOL. However, by midday the clouds had moved in, and as the skies darkened, so too did the mood in the town; it was becoming as variable as the weather.

On the one hand, hope that any more men would be found alive was waning. How could it not? Eighty-one survivors had been rescued from the shattered No. 2 mine. Another body had been retrieved from the depths, meaning eighty-two men were still missing. Despite the tireless round-the-clock efforts of mine manager Harold Gordon, more than two hundred barefaced miners, and fourteen teams of draegermen, no one had been rescued since Friday morning. "In places, the thundering bump buckled both the floor and the roof of the mine," the *Chronicle-Herald* reported. "Where the distance between the two was once 10 feet, there was now less than 10 inches."[1]

Gordon, unwilling to give up, had driven himself to exhaustion in trying to deal with this reality. Finally on Saturday night,

DOSCO's medical chief, Dr. J.G.B. Lynch, had ordered him to stand down and take a rest. As the chief organizer of rescue efforts, Gordon had been front and centre since his early Friday morning arrival in town, and he'd barely slept. The rescue crews that were working under his direction had made remarkable progress searching the coal faces at 13,400 and 13,800 feet. They'd brought survivors to the surface as quickly as possible, and they'd recovered eleven corpses so far. This latter work was becoming more difficult and horrific by the hour.

The decomposition of a dead human body begins mere minutes after death as cells are starved for oxygen, and after a day or two the internal organs start to break down. Typically, bodily fluids then start to leak from orifices, the corpse balloons, and the skin turns a greenish, mossy hue. The process isn't pretty, but it's as natural as it's inevitable.

The rate at which a body decomposes depends on a variety of factors. Not surprisingly, temperature is a key one. Deep underground in the ruins of the No. 2 mine, corpses were decomposing quickly in the ambient 80-degree-Fahrenheit heat. The resulting stench was overpowering and pervasive—so much so that it was almost impossible for the men who were recovering bodies to remove the odour from their clothing or hands. Even more problematic was that whenever a compacted body was unearthed, it immediately began to balloon. When this happened, leaking bodily fluids often splashed onto the recovery workers.

DOSCO officials contacted the army researchers in Ottawa and secured a shipment of face masks with charcoal filters. These masks might have been effective if not for the heat that fogged them up, making it virtually impossible for the men who were wearing them to see, and so they discarded them. While the work of recovering bodies from the mine was nauseating, it was the proverbial dirty job that had to be done.

Bodies removed from the mine were encased in airtight aluminum coffins. Each time one of these coffins was brought to the surface, it was transported to a garage on the mine property, and there a representative from the mining company, an official from the miners' union, and a doctor—usually Dr. Arnold Burden—opened it. Once the identity of the dead man had been confirmed, the inspectors each signed three separate tags. One stayed with the body, one was affixed to the coffin, and one was kept in the DOSCO files. If a body was unrecognizable either because it had been squashed flat or was too badly decomposed, the only means of identification was by way of distinctive body markings and dental records. It was grim work, not for the squeamish.

The bodies of miners that were recovered from the shattered mine were encased in aluminum coffins. (Ted Jolly, courtesy of Eleanor and Daniel Jolly)

The rising body count and the steady stream of funeral cortèges passing along Main Street roiled emotions and drove home with stunning clarity the Bump's true impact. Virtually everyone in Springhill knew at least one of the men who were missing and presumed dead or whose body had already been recovered. The disaster had torn a gaping hole in the town's social fabric, and it had underscored the severity of the economic challenges facing the DOSCO mine going forward.

Springhillers understandably were heartbroken. Some sought solace in their religious beliefs. Some were frustrated by the randomness of the pain and loss they were feeling. Some were downright angry. They desperately needed someone or something to blame for all the bad things that had happened to them and to their town. Such is human nature. It has often been said that the search for a scapegoat is the easiest of all hunting expeditions. That is true. And in Springhill in the wake of the Bump, there was a small group of louts who, because of their insensitivity and arrogance, unwittingly made themselves easy targets. A few journalists poisoned the well for themselves and their colleagues as a result of their boorish behaviour, which prompted many Springhillers to mistrust all journalists. While the motivations were different, the scorn for the media that's so common today is nothing new.

IT'S LIKELY THAT few, if any, of the big-city journalists who rushed to Springhill to report on the disaster had ever visited the town before—or that they'd even set foot in the province of Nova Scotia. That doubtless was true for the newshounds from central Canada, the United States, and the United Kingdom. It's not surprising that some of the journalists who parachuted into town were intent only on obtaining the spectacular details of the disaster as quickly as possible and then getting back to "civilization." A handful of these impatient intruders had little—if any—regard for the feelings of the locals.

Wally Hayes, the fledgling reporter-photographer who was a member of the *Chronicle-Herald*'s journalistic team in Springhill, witnessed some of this behaviour first-hand. Hayes felt embarrassed to the core by what he saw. Yet there was one positive: his experiences were edifying. They prompted him to resolve never to be that mercenary or insensitive while he was doing his job.

Surprisingly, it was one of the *Chronicle-Herald* team who was front and centre in that regard. The man, a veteran photographer, had a deft eye for images that captured the essence of a news story. At the same time, "[He] let nothing stand in his way of getting the photo he was after, much less the feelings of the individuals he was photographing. I found this strange because [he] was a very religious, born-again Christian," said Hayes.[2]

On one occasion, while observing the Sunday service at a local church, the photographer goaded a distraught young woman until she completely lost her composure and ran crying from the building. Ever keen to capture scenes of human drama on film, the photographer raced outside before the young woman's tearful exit, and he then snapped several photos. One of them appeared in the next day's edition of the *Chronicle-Herald*. This troubled Hayes so much that he questioned why the photographer had behaved as he had and then taken advantage of the young woman's vulnerability. That query elicited "a characteristic smirk [from the photographer] that as much as said, 'I got the picture I wanted, didn't I?'"[3]

Another time, that same pushy photographer intruded on a wake that was being held in the home of a miner who'd perished in the Bump. The parlour of the small, simple home was crowded with mourners, one of whom was a teary older woman in a wheelchair. A clergyman was doing his best to console her. In his mind's eye, the newspaper photographer pictured a scene he'd decided had the makings of a memorable photo. Approaching the clergyman, he said, "Hey, Bud! Wheel the old babe over here next to the coffin so I can make [take] a picture." Hayes, who was standing nearby, was aghast. He bolted for the door, jumped into his car, and raced back to the media room at the DOSCO headquarters on lower Main Street.[4]

Springhill being a small town, word of the incident at the wake inevitably reached Arnie Patterson, the coal company's public relations director. Patterson was outraged, and when the photographer—who'd somehow escaped the wake with his life—reappeared in the media room, Patterson pounced on him. The ensuing scuffle ended with Patterson tossing the insensitive shutterbug down a flight of stairs and out into the street. "Get the hell out of Springhill, and don't come back!" Patterson shouted.[5]

Like many of the locals, the DOSCO public relations chief was disgusted by that *Chronicle-Herald* photographer and any other journalists who were guilty of crass, insensitive behaviour. In the circumstances, that disgust was more than understandable. With the search for bump survivors dragging on, the chances that rescue workers would find any of the missing men alive were growing slimmer. The mood of the townspeople was darkening.

THERE WERE STILL two areas of the No. 2 mine that rescue crews hadn't yet been able to access: the 13,000-foot wall and the one at 12,600 feet that had been mined out and now sat largely dormant. Harold Gordon figured the latter might provide entry to the top end of the coal face at 13,000 feet. More than fifty of the missing men were believed to have been working there when the Bump hit on Thursday evening. If so, it was there that rescuers figured they would find these men—alive or dead. Gordon had hoped to reach it by gaining access from above or possibly via the level itself. Neither possibility was working out. At about 11:00 p.m., tired rescue workers had returned to the surface with disheartening news. They'd discovered that firedamp concentrations were dangerously high on both levels, and both were impossibly clogged with debris.

The work of the draegermen who were risking their own lives in hopes of saving the trapped miners was exhausting. (Clara Thomas Archives and Special Collections, York University)

Harold Gordon had decided the most promising option was to dig a new access tunnel through the several hundred feet of virgin coal and sandstone directly above the 13,000-foot level. And so rescue workers set to work carving out a three-foot by three-foot passageway that would be large enough for fully suited draegermen to crawl through.

Because the Bump had heaved trolley rails in the mine up to the roof in many places, all the wooden timbers needed to brace this new access tunnel had to be moved by hand. What's more, the men doing the digging in such a confined space had to crawl on hands and knees while wielding sawed-off picks and shovels and doing their best to manoeuvre a bulky air-powered chipper. It was that latter piece of equipment that created the staccato *rat-a-tat-tat* noise that stirred hopes of being rescued in the two groups of desperate men trapped deep underground.

Every rock, lump of coal, and clod of dirt that rescue workers removed from the new tunnel had to be laboriously hauled in buckets that were passed hand to hand, one rescue worker to the next, along the length of tunnel. The work was exhausting. It was endlessly frustrating, and it was agonizingly slow. Progress was 10 to 12 feet during each eight-hour shift. Sometimes, when the workers ran into solid rock, it was even slower.

THE FIRST FUNERALS for victims of the Bump were held on this cool, overcast Sunday afternoon. More than two hundred people—relatives, family friends, and the minister and choir members at the church Harry Halliday and his family regularly attended—jammed into the family's single-storey house on Main Street. Afterward, the funeral cortège wended its way east along Main Street to Hillside Cemetery, the windswept patch of ground east of town where Springhill miners had been buried for more than eighty years. Eva Halliday, Harry Halliday's blind mother, and his two daughters stood quietly weeping at the graveside as Halliday's coffin was lowered into the earth. Then, as one of the big-city newspaper reporters who attended the funeral would write, "The sightless woman groped in her own darkness, her trembling hands seeking to touch the coffin of her son. She was helped forward by her daughter-in-law. Tenderly the mother patted the gray box, whispering, 'Goodbye, Harry. May God bless you.'"[6]

A newspaper photographer who was lurking on the periphery of the funeral ready to capture the poignant moment on film raised his camera to snap a photo. But then he stopped, lowered his camera, and let the heartbreaking moment pass unrecorded. "It was the first time I had shed a tear since I'd arrived in Springhill," observer Wally Hayes would recall.[7]

The gravity of the disaster and its consequences for the town were starting to sink in for one and all. A profound sense of sadness and foreboding pervaded Springhill. When Dr. Noel Murphy, a psychiatric resident at Dalhousie University in Halifax, arrived in town intent on supporting those who needed help managing or dealing with their grief—in particular, the wives and mothers of missing miners—he confided to a colleague that the situation he encountered was the saddest he'd ever seen. "There are those [women] who scream and pound the walls, and those who sit and stare at the ceiling and say nothing for two or three days," he said.[8]

Families and loved ones of the missing miners didn't want to believe those eighty-two men would never be coming home again. During morning services in Springhill's nine churches, the pews were packed to overflowing with those who prayed for the safe return of their loved ones. Margaret Guthro, the wife of Hughie Guthro, who had no way of knowing her husband was one of the twelve men entombed at the bottom end of the coal face on the 13,000-foot wall, missed the service at St. Andrew's United Church. She had a good reason. The previous day she'd chanced to look out the window just as Reverend Douglas Tupper, the minister at St. Andrew's, came striding up the walkway toward her front door. Margaret naturally assumed the worst: the man of the cloth was coming to deliver bad news from the mine.

Flinging open the door, Margaret Guthro blurted, "Hughie is dead, and you've come to tell me . . ." Then she burst into tears.

Tupper was stunned. He shook his head. No, he hadn't come with news about Hughie Guthro or anyone else. He had come to deliver cookies that one of the ladies in the church congregation had baked for the Guthros. Putting his arm around his sobbing parishioner, the Reverend led her to the sofa and sat beside her while he explained the real purpose of his visit. He could only apologize for having given her such a fright. He then promised not to return until he had news to report. With that, Tupper rose and let himself out. Margaret Guthro, shaky as a withered leaf, remained on the sofa. It took her a long while to regain her composure; next morning, she was still too rattled to attend the Sunday service.

Margaret Guthro wasn't the only woman in Springhill who was racked with fear that her missing husband was dead. Far from it. Norma Ruddick was plagued with the same anguished thoughts. Still recovering after having given birth on October 14 to a baby daughter—her twelfth child—she was fearful that the challenges she already faced were about to become even greater. It was hard enough for a couple to raise a large family; it would be almost impossible for a single parent, especially a woman in her mid-thirties who had a dozen mouths to feed and bills to pay. The three eldest Ruddick girls, Colleen, Sylvia, and Valerie, were doing their best to help out, caring for their siblings and keeping up the family spirits. While Norma thanked God she had her faith to sustain her, she knew optimism and religious fervour wouldn't be enough to feed and clothe her family. Life without her husband would never be the same. Maurice was the family's sole breadwinner, and he was the focal point of life in the Ruddick house. He was even more conspicuous in his absence than he was when he was present. His chair sat empty at the head of the family table. His spirit pervaded every corner of the family's weathered old company house on Herrett Road.

Meanwhile, at the Tabor family Pleasant Street home, on the east side of town, Ruth Tabor was worried sick about her husband, Tommie. Ruth's three brothers had all been working in the mine at the time of the Bump. All had survived and had escaped serious injuries. However, there was no word on Tommie's fate. "We heard nothing about Dad for a couple of days. We just knew that he was among the missing. All we could do was pray he was all right," said Tommie's daughter Valarie.[9]

A few miles west of Springhill, at the half-built Brine home on Mountain Road, the fears Joan Brine was feeling for the safety of her husband Harold, missing and presumed dead, were tinged with anger. What she resented as much as anything was that life was so unfair. As Joan Brine would tell a newspaper reporter, she'd "been after Harold for years to get out of the mine." In the wake of the 1956 disaster, she'd become especially adamant, telling him she didn't care what he did for work, as long as he quit the mine. The steady paycheque was welcome; the daily stresses and worries weren't. They were wearing her down. "I lived in terror, expecting anything," she said.[10]

If Harold Brine listened to his wife's demurrals, it seems her words went in one ear and out the other. He was all too aware of the risks of being a coal miner. They never bothered him. He never worried that something bad would happen to him at work; it wasn't in him to do so. If Brine ever actually bothered to look for another job, he never found one. Even after the Bump left him trapped deep underground, he persisted in his *que será, será* attitude. "We thought maybe we were done," said Brine. "When we tried to dig our way out and found we couldn't, our only hope was that someone would find us."[11]

Hope has never been a good retirement plan.

CHAPTER 20

TOUGH SLUGGING

MONDAY, OCTOBER 27, 1958
6:00 A.M.

ANOTHER DAY, MORE BAD NEWS. AT 6:00 A.M., THE OFFI-cial death toll in the Springhill mine disaster rose to eighteen when seven more bodies were removed from the pit. A steady drizzle that was falling on this glum morning further dampened the spirits of the dwindling throng of people who continued to stand vigil at the pithead. Dispirited, they watched in silence as, one after another, seven aluminum coffins were loaded onto trucks and driven away. Seventy-five men were still missing.

One of those caskets contained the body of Charlie Burton, one of the overmen who'd been working underground at the time of the Bump. Charlie had paused to record in his notebook the time of the seven o'clock mini-bump that preceded the Big One. When the barefaced miners found Charlie Burton's body, his notebook was in his jacket pocket, and he was still clutching his watch; it had stopped at 8:06 p.m., the moment the Bump ended his life.

Another of the aluminum caskets brought to the surface this

morning contained the body of Tommie Tabor, husband of Ruth, father of four, keen baseball lover, WWII veteran, and a pillar of the Springhill community. Tommie had died instantly when the coal face at 13,800 feet heaved, crushing men, machines, trollies, tracks, and sundry equipment with a force that was as sudden as it was lethal.

While official confirmation of Tabor's death wasn't totally unexpected, it still came as a shock to his loved ones. Throughout the four days of their ordeal, the family had clung desperately to theirs hopes for a miracle rescue that would bring Tommie home. Each day, the *Chronicle-Herald* and the hometown *Springhill-Parrsboro Record* published lengthy lists of the names of all the men who were still missing and of those whose bodies had been recovered. Tommie Tabor's eight-year-old daughter Valarie pored over those lists, searching for her dad's name. "When one day I looked and couldn't find it, I got really excited," she recalled many years later. "Then I saw at the bottom of the page the words, 'To be continued on the next page.' When I turned the page and saw Dad's name was still listed among the missing, it was like a kick in the stomach."[1]

Young Valarie's hopes for her dad's safe return were dashed forever on this day. When her brother and uncle came looking for her, they asked her to come outside with them. "My brother said, 'They found Dad today.' I was still pretty naive, and so I said, 'Is he OK?' My uncle shook his head. 'No,' he said. 'I'm so sorry. Your Dad has been killed.' That was it, I broke down and cried like there'd be no tomorrow."[2]

Over at the No. 2 mine, Harold Gordon also felt like crying. The DOSCO manager was dispirited after having spent another day in the mine. The rescue mission wasn't going well. Work on the rescue tunnels being dug was still agonizingly slow. When Gordon finally returned from the depths and walked over to the DOSCO offices for the daily press conference that had become his routine, he appeared tired and downcast. His eyes were bloodshot, and he coughed like a two-pack-a-day smoker; the air in the mine was still thick with coal dust.

Wally Hayes, the junior person on the *Chronicle-Herald* news team, didn't normally attend Gordon's news conferences. On those occasions when he did, he invariably found them to be downers. It seemed the news was always discouraging. That was certainly so again tonight. "By the time the last count is made— and it may be a week before all the bodies are brought to the sur- face—the [death] toll will stand at 93," one newspaper reported.[3] Gordon seemed to confirm that prediction when he confided he retained little hope that any of the missing men would be found alive. "There's not much we can do but keep slugging, and it's tough slugging," he said.[4]

That was all many of the out-of-town journalists—both print and broadcast—needed to hear. Convinced the disaster story was winding down, many of them packed up and left Springhill. "As I left the town, it was raining and wisps of fog hung over it like funeral shrouds," wrote *Globe and Mail* reporter Bruce West. "Up on the hill in the belfry of the Roman Catholic Church, the bell was ringing slowly—a clang and a six-second pause—a clang and a six-second pause."[5]

Even the *Chronicle-Herald* pulled its reporters and photog- raphers back to Halifax—everyone, that is, but Wally Hayes. As the newspaper's junior man in Springhill, he found himself designated to stay behind. His job going forward was to report the names of any victims whose bodies were recovered. It was a thankless assignment, one Hayes didn't relish. He was acutely con- scious of the level of antipathy many locals felt toward members of the media, especially those who came from away. Springhill was small enough that strangers were as conspicuous as beacons. Wally Hayes learned that one day as he was sitting on a bench near the big Red Cross tent near the pithead. "Beside me were three teenage girls," said Hayes. "The girl nearest me asked where I was from and what I was doing in Springhill. I told her that I was a reporter-photographer with the *Chronicle-Herald*. She looked at me and said, 'I hate reporters.' At that, the girls got up and walked away, leaving me alone."[6]

HALF A MILE below the surface, at the bottom of the 13,000-foot coal face, the Twelve awoke to the distant and muffled sound of hammering. The noise had the impact of a jangling alarm clock. Harold Brine and the rest of the trapped men were instantly alert. In the excitement of the moment, they momentarily forgot the hunger gnawing at their bellies and the thirst that was causing their parched lips to split and making the simple act of swallowing torturous.

"Somebody is coming," a voice cried out in the darkness. "They haven't forgotten us!"

"We've gotta let them know we're alive," someone else shouted.

And so they did. The men began hooting and hollering like lunatics, summoning every ounce of energy they could still muster. They bellowed at the top of their lungs. They groped in the darkness for something, any object, that they could use to bang on the floor or wall of the mine. Harold Brine still had enough energy left to scramble to his feet and search for something with which to hammer on the end of the broken air pipe that protruded through the mine rubble and into a distant corner of the space where the men were entombed. The tumult the dozen men created continued for several minutes until gradually, one by one, they ran out of energy. Spent, they plopped back down on the mine floor, where they lay light-headed and panting. Weak from lack of food and water, the effort had left some of them feeling nauseous. Others gagged and retched.

"Damn, damn, damn!" bellowed Bowman Maddison. He was ready to cry and might have if he'd had any tears left in his dry eyes. "They didn't hear us."

Now that the uproar they'd raised had ended, the men realized to their dismay that the hammering that had so briefly raised their spirits had ended. Once again, there was only a gaping, empty silence punctuated from time to time by the sound of one of the Twelve coughing, moaning, or weeping. With such a void crying out to be filled, Caleb Rushton began to sing a hymn. He started slowly, the sound welling up from somewhere deep within him.

Rushton had sung for many years with the choir at the Anglican church he attended every Sunday. On this mournful day, four days after the Bump, he could have spat dust; his mouth was that dry. Yet somehow the volume of his voice, creaky at first, gradually rose until it filled the chamber with a soulful sound that lifted the spirits of the other men. They soon joined in the singing. Even Harold Brine, who was a lapsed Catholic and not much of a singer. The great French writer Victor Hugo once observed that "music expresses that which cannot be said and on which it is impossible to be silent." At that moment, any one of the Twelve who raised his voice in song would have agreed with M. Hugo.

When the hymn ended and the music died away, that painful silence returned once more. Presently, it was the ever-upbeat Bowman Maddison who broke it. "I once heard that a bit of coal will cure heartburn," he said. "I guess that means it's all right to eat it. I'm awfully hungry, so I think I'm gonna give it a try."

With that, he scooped up a chunk of loose coal from the floor of the mine and popped it into his mouth. He didn't mention how awful it tasted. Instead, he munched away in the darkness. To the men listening in the silence, he sounded like a rat chewing on a carrot.

"Not bad," he said. "And it's sure a lot better than nothing."

Even though the others didn't believe him, one by one, they too picked up tiny bits of coal and started to nibble. For his part, it took Harold Brine only a few small bites to decide that was more than enough for him; if he ever got out of there, coal wasn't something he'd ever want to have for supper.

SUSTENANCE ALSO HEADED the priorities list for the men trapped at the top end of the 13,000-foot coal face. After another agonizing night of moaning, crying, and calling out incoherently, Percy Rector was finally quiet, for a few minutes, at least. His companions in the darkness welcomed the respite however brief it might be. It gave them a few much-coveted minutes of silence in which to sleep or to dream of something other than how ravenously hungry and thirsty they were.

Dreaming. That's what Maurice Ruddick was doing. He was so hungry, his stomach ached. He'd have given the world for a bowl of his wife Norma's stew or a slab of that delicious bread she made fresh each morning.

In Springhill at this time of year, sunrise was a few minutes after 7:30 a.m., and so Ruddick knew it would still be dark up above. In the pre-dawn, all would be quiet at home. Most of the kids would still be fast sleep. If wife Norma was out of bed—and she usually was—Maurice could imagine her in the kitchen, where she'd be baking bread or supervising her eldest girls while they rolled dough and fired up the big old cook stove. Maurice's singing "Minerettes" were fledgling bakers too.

To help him forget food and home, which was barely a half-mile above the spot where he and his companions were entombed, Ruddick lay there staring at the luminous dial of his watch. He was lost in his reveries when suddenly he realized someone was calling to him. Ruddick recognized the voice of Garnet Clarke.

"*Pssst! Pssst!* Maurice," he whispered. "What time is it?"

"It's still early," said Ruddick.

"Maybe it is, but it's October 27. Today is my birthday," Clarke said.

"Is that right? Well, happy birthday, Garnie," said Ruddick. "How old are you?"

"Twenty-nine," came the reply. "The way things are looking, I guess I won't see thirty. This will be my last birthday."

"I always take the day off when it's my birthday," said Ruddick, changing the subject. "Today, I think you should too, and we can celebrate with you."

With that in mind, Ruddick began calling out. He wanted to rouse everyone so he could share the good news that today was Garnet Clarke's twenty-ninth birthday. Time for a party, Ruddick suggested. While they had no birthday cake, they still had some water and one crusty sandwich. They could pretend the water was beer and the sandwich was a birthday cake. Why not?

Before the party could begin, Ruddick had one vital chore to do. He broke a tiny piece off the corner of the bread and took it to Percy Rector, who was drifting in and out of consciousness. Ruddick had become Rector's caregiver by default. He'd known Percy and worked with him for more than a decade. Now in Percy's hours of greatest need, it was Ruddick who stepped up to comfort and protect him at a time when he had little, if any, awareness of what was happening around him. Despite this, Rector somehow clung to life. This morning, when Ruddick pressed the tiny bite of sandwich to the poor man's lips and followed up with a few precious drops from the water can, Rector swallowed and cried out for more. There wasn't anything more to give. He'd had his share. The rest of the water and the sandwich were for Garnet Clarke's birthday party.

Ruddick broke the remnants of the stale sandwich into pieces. Each of the men crawled over to him through the darkness. It was like some bizarre sort of miners' communion, with Ruddick in the role of priest. He pressed a morsel of bread between the parched lips of each man. And after doing so, Ruddick offered up the water can, which by now was almost empty. There was barely enough liquid left for each man to wet his lips, let alone taste the water. When the can was finally returned to Ruddick after the last man had had a sip, it was empty. He gave it a shake and then dropped it. The jarring sound of the tin can clattering across the mine floor had a definitiveness to it.

"Well, that's it, boys," said Ruddick. "That was our last sandwich, and all our water is gone. We're out of everything now."

No sooner had those words left his lips than Ruddick regretted having uttered them. He sensed he'd spoiled Garnet Clarke's birthday party. It was in hopes of making amends that Ruddick did the only thing he could think of and what came naturally to him. He began to sing "Happy Birthday." The others quickly joined in, everybody other than Percy Rector, who was in no condition to sing, and Garnet Clarke, who figured it wouldn't be right to sing "Happy Birthday" to himself, even if this was probably going to be his final one.

CHAPTER 21

"NO ONE IS COMING FOR US . . ."

TUESDAY, OCTOBER 28, 1958
6:00 A.M.

IN THE BIG SALVATION ARMY TENT THAT SAT NEAR THE PIT-head of the No. 2 mine, a handful of bleary-eyed people were huddled around the pot-bellied stove. It was another cold, rainy morning in Springhill. The fire's radiant warmth and the aroma of freshly brewed coffee were powerful draws for the dwindling number of townspeople who persisted in their mine vigil and for the exhausted rescue workers who dropped by the tent after completing their shifts.

Seven funerals had been held in Springhill the previous day; six more were planned for today. The disaster's official death toll now stood at twenty-two, and no one doubted that number would rise, especially after what a grim-faced Harold Gordon would have to say when he returned to the surface after another long, gruelling night down below. At 9:00 a.m., the DOSCO manager held his daily press conference in the company offices. Once again, the news he had to share was deflating. It was a repeat of the message he'd delivered the previous day. "The rescue crews are making only ten-to-twelve feet in each eight-hour shift. They can't go through any faster. It's almost solid rock ahead of them," he said.[1]

Asked by a journalist to clarify whether he was saying all the miners who were still missing were dead, Gordon replied, "There's no reason to think otherwise."

When word of Gordon's pronouncement reached the pithead, people there felt devastated. Suddenly, it felt as if all hope had ended. "Word of Mr. Gordon's announcement has slapped through this crowd like wildfire, and it's put a sort of silence, a pall, over the people who have waited since last night," Jack MacAndrew of CBC reported in a radio newscast. "Some of them are still waiting, but more are leaving in ones and twos and threes. They hunch their shoulders against this driving wind and icy rain it's carrying, and they're going uptown back to their homes."[2]

In the wash house, those rescuers who were donning their gear and getting ready to go below for another shift were also taken aback. Many of these men were pragmatic, no-nonsense guys who weren't shy about venting their anger; they cussed Gordon roundly. Others simply refused to accept—let alone believe—their rescue efforts were now a recovery mission simply because the boss said as much. Gordon's words didn't matter. He might be giving up; the draegermen and barefaced miners wouldn't. They refused to do so. They were determined to press ahead with their rescue mission. Always in the back of each man's mind were the tenets of that unwritten miners' code: any miner who's missing is still alive until his body is found, and no miner is ever left underground even then.

From one side of town to the other, people reacted to Harold Gordon's pronouncement with a mixture of sadness and resignation. Tuesday marked five days since the Bump, and none of the miners who'd been working the afternoon shift had emerged from the mine alive since the previous Friday morning. And so it was with good reason that people feared it was only a matter of time until DOSCO confirmed what no one wanted to say: all the miners who were still missing were dead.

The implications of that grave assumption were underscored by developments that began happening all over town. For one thing, DOSCO officials announced that the widows of all those men whose bodies had already been recovered would start receiving survivor benefit pension cheques later that same week. That decision wasn't altruistic in any way; the coal company was legally obligated to provide survivor pensions when miners died on the job. If that hadn't been the case, it's unlikely the coal company would have treated its employees with such benevolence. The attitude of DOSCO's corporate owners in far-off London made the penny-pinching Ebenezer Scrooge seem like a benevolent employer by comparison. The moment the earth shook and coal mining came to a halt—at 8:06 p.m. on the evening of October 23—the coal company had stopped paying the men who were at work in the No. 2 mine. Remember the company rule: no coal, no pay. That was how it was; unbelievably, there were no exceptions. "Boys, you're on your own time now," DOSCO officials seemed to say.

LOST WAGES WERE the last thing on the minds of the men trapped deep within the mine. Their only priority was survival. After having gone so long without food, the men were hungry, desperately so. They were starting to feel weak and were experiencing stomach pains; when you haven't eaten for several days, the stomach acid that normally dissolves food starts to attack the lining of your gut.

Unpleasant though that is, for anyone in dire straits a lack of water is a far more pressing and critical problem than a lack of food. The body of an average adult male is about sixty percent water (the comparable percentage for a female is slightly less);

that's roughly 45 quarts. To stay topped up, a man typically needs about 15.5 cups of water per day, a woman 11.5 cups.

With an adequate supply of drinking water, you can survive for several weeks without food. The survival period without water is much shorter; serious health issues start cropping up after just four or five days. A starving human body can break down its own tissue to replace the nutrients it normally derives from food, but it has no way to deal with dehydration.

Bowman Maddison's suggestion that he and his eleven companions ease their hunger pangs—and pains—by eating coal may have sounded at first blush like a reasonable idea to starving men who were desperate for something to eat. But it wasn't. The simple fact is that the human body can't digest coal. Ingesting it will cause not only stomach problems and possibly even physical damage, but also constipation.

Not knowing this, on Tuesday—his fifth day without any real food—Bowman Maddison continued to munch on small bits of coal. His companions in misery had pretty much given up doing so even though hunger was gnawing at their insides. Compounding their woes, the lack of drinking water had become life threatening. Dehydration was already making them feel weak and disoriented.

As he sat there in the dark, Harold Brine reflected that even with a watch on his wrist, it was becoming ever more difficult for him to remember what day of the week it was. Mind you, what difference did it really make? The calendar has no meaning when you're three-quarters of a mile underground, surrounded by perpetual darkness, starving, and dehydrated. Despite that, some of the men continued counting days.

"What day is it?" Levi Milley suddenly asked.

"Tuesday, I think," came the reply from Caleb Rushton. "Is that right, Harold?"

"Yeah. It's Tuesday," said Brine.

"That means we've been here five days now. Five days," said Milley. "The gas is probably bad, and if it is, they've likely sealed off the mine, like they did in '56. If that's what's happened, we're done for. All of us. We'll die here!"

The words were no sooner out of his mouth than, as if on cue, from far in the distance came the muffled sound of more hammering. His hopes having been rekindled, Bowman Maddison scrambled the ten paces over to the broken air pipe. After frantically groping around in the darkness, his hand hit upon a water can. Snatching it up, he began banging on the pipe. He did so for an hour, until exhaustion finally overcame him and he collapsed on the mine floor. In the distance, the hammering had ended, and now there was only that tormenting silence again. The disappointment was soul crushing.

FOUR HUNDRED FEET away, on the other side of the wall of debris that clogged the head between the 13,000- and 12,600-foot levels of the shattered mine, Maurice Ruddick was lost in thought yet again as another day began. Day Five of their entombment. It was the wee hours of the morning, and to pass the time he was busy trying to compose a song about the No. 2 mine: "The 23rd of October, we'll remember that day / Down the shaft underground in our usual way . . ."[3] Ruddick imagined himself singing those words into the microphone of the old reel-to-reel tape recorder that sat atop the rickety wooden table in the backroom at home. Oh, what he wouldn't give to be at home right now.

Coming up with a tune was Ruddick's way of diverting his mind from thoughts of food, water, and the horror of his predicament. Then too, there were the hideous, unrelenting realities of Percy Rector's ordeal; there was no escaping them. From time to time in his delirium, Rector threw what little energy he could still muster into a fruitless effort to tear off the now useless arm that gripped him with so much pain and misery. Even in the pitch black, the other men could tell what was happening. Percy moaned and wailed. In his confusion, he cried out for family members. He called out to his kids. He issued commands to his horse, and over and over he pleaded for the water that no one there could give him. In his rare moments of lucidity, Rector repeated his plaintive cries for someone, anyone, to hack off his arm. "For God's sake . . . won't someone please help me? *Please!*" he wailed. Such frenzied cries, which were anguished, guttural, and fearsome, were soul searing. It was all enough to make a man scream or cry. Or both.

"There was one fellow that did grab the axe and was thinking about it because he was all worked up," Frank Hunter said. "But I stopped him. . . . I said, 'There's such a thing that somebody might come along and get us out and they might save him.'"[4]

And so it was that Rector's pleas continued to go unheeded. No one was willing to do what Rector was pleading for. Even if one of the men had taken it upon himself to carry out the amputation, doing so in the pitch black would have been all but impossible. Percy's only hope, and it was slim at best, was that rescuers would arrive. The possibility of that happening was rekindled yet again in a flash the instant the men heard the distant sound of hammering.

They'd just finished a dietary experiment. Garnet Clarke had hit upon the idea that a breakfast of tree bark might help ward off starvation. With that in mind, he'd stripped some bark off a chunk of pack timber, spruce wood, he had found in a corner of the chamber. After gingerly tasting a small sample, he popped the rest of the piece into his mouth and began chewing like a dog with a stick. Maurice Ruddick decided to give it a try too. When he did, he was surprised.

"You know, that doesn't taste half bad," said Ruddick. He might well have added the qualifier, "Or as bad as I thought it would." But he didn't bother.

Soon, all the men were like beavers nibbling on strips of tree bark. Afterward, they were sitting in the dark, literally waiting to see how their ersatz meal would go down. It was at that moment that in the distance they again heard thumping and hammering. This time, there seemed to be something different about the noise. For one thing, it sounded as though someone was tapping on the far end of a broken metal air pipe that lay over in a corner of the chamber. As he sat there straining his ears to listen, Maurice Ruddick counted six taps. He wondered if they were a signal.

Garnet Clarke wondered the same thing. He scrambled through the darkness, snatched up an empty water can he managed to find, and set about banging on the air pipe. He did it six times in reply to the taps he'd heard. It was with bated breath that Clarke, Ruddick, and every other man in the chamber waited for a response. When it came . . . there were six taps. Any elation the men felt was short-lived, for soon there was a seventh tap. Then an eighth, and then five more followed.

"That's thirteen taps!" a voice said in the darkness.

"Damn it! They didn't hear my tapping," Clarke cried.

The sense of disappointment in the chamber was profound. No one spoke afterward for the longest time. It was deathly quiet. Lying there in the wake of this latest setback, Maurice Ruddick let his thoughts drift far away again when something caused him to become aware of Percy Rector's breathing. The sound and the pattern had changed. Ruddick sat up immediately and clambered over to check on Rector. When he reached out, his hand touched one of Rector's legs. To Ruddick's horror, it felt cold. A moment later, Percy Rector took one last gasping breath, and then he fell silent. It was 4:20 p.m. when the life went out of him. His suffering had finally ended, and so too had his piteous cries.

Maurice Ruddick could barely speak. "Boys . . . Percy's gone . . ." he said, his voice breaking.

In the silence that followed, each of the men had the same thought. With death now there among them, each man in this dark, foul hellhole wondered how much longer before his own suffering would come to an end.

"I was just laying [*sic*] there, waiting to die or to be rescued, one or other," said Herb Pepperdine. "Nobody was saying a word."⁵

As the life-and-death drama played out deep within the shattered No. 2 Springhill mine, scores of volunteers from the Red Cross and other organizations waited patiently above ground while continuing their tireless work in support of the rescue and recovery operations. (Ted Jolly, courtesy of Eleanor and Daniel Jolly)

CHAPTER 22

SORROW, ANGER,
AND DESPERATION

WEDNESDAY, OCTOBER 29, 1958
7:00 A.M.

THE NEWS THAT DOMINATED PAGE ONE OF THE DAY'S EDI-
tion of the *Chronicle-Herald* spoke volumes about the attitude of
Springhill residents and the media toward the town's latest mine
disaster. "Thousands Cheer New Pope," announced the bold-type
headline that appeared above photos and a report from Rome providing details of the election of Pope John XXIII. At the top of the
page, a smaller subhead carried the ominous news "DOSCO Says
Springhill Mine Future Undecided."

It's true what they say: hope does spring eternal. However, six
days after the killer bump, almost no one in Springhill still believed
that any of the sixty-seven men who remained missing would be
found alive. That was true even for the wives and children of the
missing miners and for the three teams of rescue workers who continued to doggedly persist in their round-the-clock efforts to dig a
rescue tunnel.

Bowman Maddison's wife, Solange, had called her life insurance company to give notice that Bowman was missing and presumed dead; she needed any lump sum the policy, which was just three months old, would pay. Hopefully, it would cover the cost of a funeral. When a company agent called from Moncton, he advised Mrs. Maddison that a cheque for $6,000 would be coming her way just as soon as her husband's body was recovered and had been officially identified.

Margaret Guthro, like Solange Maddison, had resigned herself to the notion that she was now a widow. With that in mind, she set to work planning her husband's funeral. Hughie Guthro, like Harold Brine and so many other Springhill miners, had built his family home with his own hands. And like Harold Brine's place, after eight years the Guthro residence was still a work in progress; Hughie had never gotten around to building front steps. Whenever Margaret had asked him about it, he'd insisted that he'd get to work on the steps "tomorrow." That day had never come, and now it might never come. The problem was that without front steps, there was no way for pallbearers to carry Hughie's casket into the living room. Distressed by that thought, Margaret called her brother to ask for help. He volunteered to build a set of front steps for the house.

On Mountain Road, Harold Brine's wife Joan had long since given up any idea of going to the mine and joining Harold's sister Lorraine in the vigil she and so many other women, children, and older men were maintaining. Joan couldn't bear to be there, even when she found someone to sit with her daughter. She preferred to stay home and hear any bad news on television.

Over at the DOSCO offices on Main Street, the mood was similarly downcast. The press room where Harold Gordon had continued to hold his daily press conferences was quiet just now. Only a handful of the journalists who'd been in town a few days earlier were still around; all the others had hightailed it out of town as soon as it seemed there would be no miraculous rescues. "There were only a few of us left," said Wally Hayes. "Myself representing the *Chronicle Herald*; Jack MacAndrew, the CBC newsman who bounced back and forth between doing radio and television coverage; and Ian Donaldson and Joe Dupuis, both reporter-editors with the Canadian Press (CP)."[1]

The building that once housed the DOSCO offices in Springhill is today the home of the local seniors' centre. (Ken Cuthbertson)

The four reporters had experienced a frightening ordeal the previous night when some young miners accosted them. Maybe it was the endlessly depressing news coming out of the mine that riled the miners, or maybe it was the persistent rumours that regardless of whether any more men made it out alive, DOSCO's management had decided to close the colliery. If that happened, hundreds of jobs would disappear instantly, and Springhill would become a ghost town. Whatever it was that got the miners riled up, they were in a foul mood.

The first indication of that mood came when an anonymous caller rang the DOSCO headquarters with a warning: a mob of angry miners was on the way over there to "clean out" the press room and run any reporters still there out of town on a rail. A fearful Ian Donaldson of CP immediately called the nearest Royal Canadian Mounted Police (RCMP) office, which was in nearby Amherst. When the RCMP dispatcher asked if any angry miners had appeared yet, Donaldson said no. "Okay, then, call back if they do," he was told.

Not long afterward, a mob of miners with malice in mind did show up outside the DOSCO offices. By now, it was dark. There was no way for the journalists to tell how many men were out there, and so Donaldson hurriedly rang the RCMP again. This time he was told a police car would come "as soon as possible." That was no comfort to Donaldson, Hayes, Dupuis, or MacAndrew. The foursome were terrified the miners were intent on coming into the building after them. Fortunately for all concerned, some quick thinking by Donaldson saved the day. He remembered that William "Bull" Marsh, the head of the mine workers' union, was in town and hastily got him on the phone; Donaldson had met with Marsh a few days earlier at the office of the Springhill local. The union head's nickname was well earned. He was as wide as he was tall, and his approach to dealing with problems was direct. "[Marsh] arrived on the run, strode into the crowd pushing bodies left and right, and in a matter of minutes the crowd had dispersed," Hayes recalled.[2]

As it turned out, most of the hotheaded miners who'd marched to the DOSCO offices were young men who hadn't been called upon to work in the rescue operation. They were also frustrated and angry that the coal company was planning to close the Springhill mine. They wanted to lash out. It didn't matter that the media had nothing to do with any such decision. The journalists who reported the news were at hand; DOSCO's senior managers seemed far away and out of reach.

JUST AS FRAYED tempers and frustrations were bubbling away on the surface, tensions were also starting to stir down below. At the top end of the 13,000-foot coal face, Maurice Ruddick and his companions were literally dying of thirst. After six long days without drinking water, all of them were suffering horribly. Ruddick's lips were cracked and swollen; it took no small amount of effort for him to even talk. There was nothing much that needed saying now anyway. What more was there to say that was of any real importance? Better, he decided, to save any voice he had left for singing the hymns that brought him and the other men so much solace. That's what Ruddick was thinking, and that's what he did as he lay there staring at the watch his wife, Norma, had given him. The timepiece was a source of comfort, even after a week in the darkness when its luminous dial no longer glowed as brightly as it once had. Ruddick's eyes now strained to see the time. Sitting there trying to do so, he was as quiet as stone. Others around were more talkative. Doug Jewkes, for one.

Jewkes, with his long, mournful face, reminded some people of a toothy lookalike for Ray Bolger, who played the scarecrow in *The Wizard of Oz* movie. Like the scarecrow, Jewkes loved to talk, and he was in a talking mood this day. He was prattling on about a topic that made everyone around him see red. Jewkes didn't have a watch to look at, and so with nothing else to occupy his mind, he sat for hour after hour in the dark thinking about how thirsty he was. As he did so, he developed an inexplicable craving for the lemon-lime soft drink 7Up. Jewkes couldn't stop talking about how much he wanted 7Up. He admitted he had no idea why—all he knew for certain was that his desire for the beverage was driving him insane. And he in turn was driving crazy those who had to listen to him go on and on about his sudden fixation. Several of the men were out of patience and were becoming angry. In the circumstances, even the smallest irritation seemed like a reason to commit murder.

"For Christ's sakc, shut the hell up, Dougie!" someone barked.

"Yeah! Put a cork in it, or I'll come over there and . . . " another voice in the darkness cried out.

"Leave Dougie be," said Maurice Ruddick. "We're all tired."

He was right about that. All of them, including Ruddick, were at their breaking points.

Garnet Clarke had sat listening to Jewkes's foolishness and to the exchange that temporarily quieted him. The more Clarke heard, the thirstier he was. That's what finally prompted him to share an idea he'd been mulling over for several days.

"Boys, Dougie is right. We all need something to drink. We haven't got any water, and we sure as hell don't have any 7Up, but we've got fluid in our bodies that looks like 7Up. And right now, it's going to waste," said Clarke. "We could dip some tree bark into it. At least give it a try. What do you say? Shall we give it a try?"

To his dismay, Clarke's question elicited no response. Frustrated by this, he muttered to himself for a few moments. Then he took action. Clarke began rummaging around until he found what he was looking for: one of the empty tin water cans. There then followed the sound of liquid trickling into the can.

You might think, as Clarke did, that drinking your own pee could save you from dying of dehydration if you're out of water and struggling to stay alive. At first thought, that might seem like a reasonable suggestion. The reality is that it's not. Drinking your own urine won't rehydrate you. If you're already dehydrated, doing so will make things worse. To understand why, you only have to remember that urine is your body's liquid waste. While urine is ninety-five percent water, it also contains a devil's cocktail of dissolved salts, minerals, and trace amounts of toxins that your liver produces. None of them are good for you. The more dehydrated you are, the higher the concentrations of these pollutants your pee contains. Mind you, because one of the components is ammonia—which has some uses as a purifying agent—the ancient Romans sometimes used urine as a mouthwash, brushed their teeth with it, and added it to their laundry water. Strange, but true. The Romans obviously had their own ideas about what smells nice and tastes palatable.

Not surprisingly, none of the Seven knew any more about the chemical content of urine or its historic uses than they'd known about how to ease Percy Rector's pain or amputate his arm. Thus, over the next hour or so, one by one, each of the men who'd been listening to Garnet Clarke found an empty water can and followed his lead. No one talked about doing so, not even the usually voluble Doug Jewkes.

THE SEVEN WEREN'T the only ones desperate enough to try drinking their own urine. At the other end of the 13,000-foot coal face, Harold Brine and the eleven men who were trapped there with him were also dying of thirst. Here it was Gorley Kempt who suggested they try drinking pee, and he was the first to actually do so. Then Harold Brine gave it a try. "It was right salty," said Brine. "But if we wanted to survive, we figured we'd have to drink it."

At first, they tried softening the harshness of the taste by inserting into the water cans strips of tree bark they'd peeled off pack timbers. When that didn't have the desired effect, they tried adding to the urine some of the Tums antacid tablets Larry Leadbetter had found in the pocket of his work pants.

"We thought it would kill the taste of [the urine]," Harold Brine would explain. "After doing this a couple of times, it just got so that it tasted sour. I suppose the salt was going into our systems, and the salt was coming out all the time."[3]

The unavoidable reality was that after five days with insufficient water to drink, the men's bodily organs were starting to shut down. That was especially true of the brain. It was becoming harder and harder to know what was real and what wasn't. Yet to a man, they knew their time was quickly running out. "Some of the boys said, 'The wife said we shouldn't be down here,'" said Brine. "Well, pretty near everyone's wives said the same thing."

Lying there in the pitch black, ravenously hungry, his mouth parched, and inhaling air that was fouled with the nauseating smell of decaying bodies, Brine knew that death would soon come for them all. His wife Joan had been right; he *should* have quit the mine.

CHAPTER 23

"THE MIRACLE AT SPRINGHILL"

WEDNESDAY, OCTOBER 29, 1958
11:00 A.M.

DOUBTING IS EASY. BELIEVING CAN BE HARD. "SEEING IS not always believing," the American civil rights leader Martin Luther King Jr. once said. The Springhill mine rescue workers subscribed to that same philosophy, and so they doggedly carried on despite the skepticism of DOSCO officials, the media, and even many members of the Springhill community doubted that anyone could still be alive in the No. 2 mine. No coal miner had ever survived being trapped underground for six days, and that's how long it had been since the Bump. No matter. The volunteer rescue workers refused to give up, either in their search for bodies or in their efforts to dig a tunnel to reach the coal face at 13,000 feet. It was there that as many as fifty-five of the sixty-seven miners still missing were thought to have been working when on the evening shift.

As midday approached, the rescue workers figured they still had about 60 feet to go before they'd reach their goal. As it happened, they were wrong; the distance was actually closer to 90 feet. And progress continued to be agonizingly slow. The men doing the digging had to lie on their stomachs, their headlamps bobbing like fireflies in the darkness. The air was stiflingly hot and reeked of death. The lead man, the one whose turn it was to lead the way in the digging of the rescue tunnel, chipped bits of coal and rock into buckets. Whenever a bucket was full, he passed it back to a man behind him. That man, in turn, passed the bucket to a man behind him. And so on it went, like a manual assembly line. Some of the rescue workers spent up to twelve hours each day underground. The work, which was hot, dirty, and exhausting, was also dangerous. The possibility of another bump was ever present, as was the threat of another cloud of firedamp.

Earl Wood, a veteran miner who'd spent the last five days as a barefaced rescuer, was wielding his pick and shovel at the head of the line this morning. Wood, who was acutely aware of the dangers he faced, immediately stopped and then retreated as a portion of the wall of coal and rock in front of him suddenly crumbled. When it did, a thick cloud of coal dust and a rush of hot, stale air filled the tunnel. Fearful of the toxic gas that might follow, Wood and the dozen men directly in the line behind him scrambled to retreat. They wriggled back to an open space on the 13,400-foot level and lay down there, as close to the mine floor as rugs, while they waited for the air to clear. When finally it did, Wood crept forward again to investigate. What he now saw grabbed his attention.

There, protruding through the debris ahead of him, was the end of a 6-inch metal air pipe that the Bump had snapped off. Wood crawled up to take a closer look and was studying it when he thought he heard someone speak to him.

"What's that? Did you say something?" Wood asked the man behind him.

"No, I didn't say anything," came the reply.

When Wood inquired if anyone else farther down the line had called to him, he was told no. Satisfied that he'd been mistaken, Wood shrugged. He then forgot about the pipe as he retreated out to the level. He and the other rescue workers waited there for the arrival of one of the coal company's safety officers, who would test the air quality in the rescue tunnel.

While all this was happening, Harold Brine was sitting in the pitch black listening as if his life depended on it, which in fact it did. Brine was sure he'd heard a noise, although he had no idea what it was. The only thing he knew for certain was that whatever he'd heard didn't sound like rats or any of the other ambient underground noises he'd grown accustomed to hearing in the last six days. Unbeknownst to Brine, Gorley Kempt had been listening too.

"Harold, are you there?" asked Kempt.

"Where else would I be? Yeah, I'm here," said Brine. "But *shhhh* . . . I'm listening. I thought I heard something strange."

"Well, I heard it too."

"Good, because I thought I might be imagining it," said Brine. "What are you thinking it was?"

"I don't know, but I'm wondering if whatever noise we heard came from that broken air pipe over there."

"Maybe," said Brine. "Why don't we crawl over and have a listen?"

That seemed like a good suggestion. The two men scrambled over to the corner of the space where the broken air pipe was. They'd already visited it several times and had even chiselled some bolts off a valve in order to gain access to the open length of pipe. From time to time, coming down that conduit from somewhere distant, was the noise of what sounded like a motor running and a lot of thumping and pounding. Sometimes it sounded as if metal pipes were being dropped or moved around. At one point, Brine and Kempt spent almost an entire day sprawled on the mine floor next to the air pipe, listening to whatever sounds they could hear. "I know what they're doing out there. They're setting up a new loader, and they're going to clean out the level. They figure we're all gone in here," Brine had opined.[1]

His pessimistic comment drew angry responses from some of the men. After that, Brine checked his tongue. He was now careful about offering his opinions on the meaning of the sounds they heard coming through the air pipe. It didn't do any good either to raise false hopes or to dash them. However, today Brine became so excited that he forgot himself. He and Kempt again crept close to the broken air pipe. "We laid there, side by side. My face was on the bottom, and Gorley's face was on top of mine with our noses stuck in the pipe," said Brine.[2]

For the next ninety minutes, the two men lay on their stomachs staring into the blackness and listening. The longer they waited in vain, the more they wondered if maybe they'd been mistaken about having heard something.

IT WAS A FEW MINUTES before two o'clock when Blair Phillips, chief mine surveyor for the No. 2 mine, arrived at the end of the rescue tunnel. He was there to check for the presence of firedamp. To do so, he inserted into the end of the broken air pipe a glass bottle filled with reactive chemicals. If there was a dangerous level of toxic gas in the pipe and in the debris field ahead, the chemicals would change colour, and the rescue crew would return to the surface until it was safe for them to resume digging.

Phillips was struggling to position the test bottle in just the right place. After fidgeting for a few moments, he leaned in so his headlamp illuminated the end of the pipe. When he did so, a glint of light dashed the length of the conduit and winked in the total darkness where Harold Brine and Gorley Kempt were watching and waiting.

"Did you see that?" said Brine.

"Yeah. I think I saw a flash of light coming out of the air pipe," said Kempt.

The two men could barely contain their excitement. Scrambling to within inches of the broken pipe, Kempt listened hard. "Voices. I can hear voices!" he cried. "Thank God. We're saved!"

"Yell into the pipe. Tell them we're in here," said Brine.

By now the other ten men in the chamber had sensed the excitement. Sitting up, they peered into the darkness toward the spot where Brine and Kempt were lying on their stomachs with their faces pressed within inches of that broken air pipe. Joe McDonald with his badly broken leg was the only one who didn't stir. But even he was listening to every sound, his dying hopes of being rescued suddenly rekindled.

"Hey, we're here! There are twelve of us. We're still alive," Kempt shouted into the pipe. "Come get us out of here. Please! Save us!"

At the other end of the air pipe, Blair Phillips could scarcely believe his ears. He was stunned. His pulse was racing, sweat trickled down his brow, and he could feel the excitement rising in his chest.

"Who are you?" he shouted down the pipe.

From the distance came the muffled reply. "I'm Gorley Kempt."

"Who else is in there with you?" asked Phillips.

"There are twelve of us," shouted Harold Brine.

Blair Phillips leaned in so he could put his lips close to the pipe. "Stay there," he called. "We'll come to get you out as fast as we can."

Those words put giddy, ear-to-ear smiles on the faces of Harold Brine, Gorley Kempt, and the other ten men with them. After more than six days of despair, they finally had the feeling they'd see their loved ones again. Brine couldn't help himself—he chuckled at Phillips's request that they stay put. They'd do that for sure; after all, where did the rescuer think they might go?

After sending one of the rescue crew up to the 7,800-foot level, where there was a telephone that could be used to let the world know a dozen men had been found alive, Phillips shouted through the air pipe again.

"What are your names?" he called.

When Brine and Kempt had finished calling out each of the twelve names, Kempt quickly added a request. "How about some water?" he said. "Can you send us some? We're dying of thirst down here."

They had to wait a while longer for that water. Blair Phillips was so keyed up that he forgot all about it when he raced off to the surface to share the names of the twelve survivors. When he was gone, Earl Wood and some of the other barefaced rescuers began shouting their greetings down the air pipe. Among them was Percy Weatherbee, who was Gorley Kempt's nephew by marriage. So overcome with emotions were the two men that it was all they could do to keep it together long enough to exchange greetings. The same was true for Harold Brine and rescuer George Hodges, who was Joan Brine's uncle. For only the second time in the almost seven days of his harrowing ordeal, Brine began to sob. He could scarcely believe that he might be saved.

AT THE 7,800-FOOT level of the mine, a breathless Blair Phillips snatched up the telephone and called the pithead. He was eager to share the joyous news that a dozen men had been found alive down at 13,400 feet. The moment the manager who answered the phone heard what Phillips had to say, he threw open the office door and began shouting like a town crier. "There are men alive down below!" he bellowed. "They've found men alive!"

That news spread with lightning speed to every corner in Springhill. Church bells began to chime. Cars with horns blaring raced up and down Main Street. Shopkeepers locked their doors, and offices emptied out as clerks joined the rush to the pithead. At the DOSCO offices, Harold Gordon was in the middle of his daily press conference when one of his staff entered the room and whispered into his ear. In the next instant, Gordon was out the door, hotfooting it for the mine.

It was almost 4:00 p.m. when Wally Hayes, the young *Chronicle-Herald* journalist, heard the same news that had sent Gordon sprinting to the mine and set the town buzzing. Hayes was passing time in the Salvation Army tent near the pithead. There were free sandwiches and coffee to be had there and people to talk to, and so Hayes had decided this was as good a spot as any to wait for announcements of the names of the miners whose bodies were brought to the surface that day. Reporting those details was the thankless job he'd inherited when the rest of the *Chronicle-Herald* news team packed up and headed home. Hayes had been told if there was any breaking news to report, he was to immediately call Halifax radio station CHNS, which was owned by the *Chronicle-Herald*, and then to call the newspaper.

Hayes had just finished a coffee when there was a commotion outside the tent and people ran off to see what was happening. When he moved to join them, one of the Salvation Army workers, an excited "spit of a woman," grabbed him by the sleeve. "The miners are coming! The miners are coming!" she shouted. After his experiences the night before, the first thought that popped into Hayes's mind was that the miners who'd marched to the DOSCO offices to throttle any reporters they could get their hands on were coming again, and this time he was on his own. Thinking he'd be wise to stay out of sight, he remained in the tent until the place was empty. By now, however, curiosity was getting the better of him. When he crept over to the doorway and peeked out, what he saw astounded him. A stream of people—young and old, men and women, Salvation Army officers and Red Cross workers—were running toward the pithead. "A man wearing a miner's helmet was passing me, and I shouted, 'What going on?' He said, 'We've found twelve miners alive,' and his voice trailed off as he disappeared into the sea of people."³

Without waiting another moment, Hayes sprinted the 200 yards to the DOSCO building on Main Street and then bounded up the stairs to the press room where Ian Donaldson, Joe Dupuis, Jack MacAndrew, and one or two other people were sitting around. They somehow weren't aware of the news that had come out of the mine, and so they shouted at Hayes to ask him what was happening. Hayes shrugged as if he didn't know. He wanted the scoop for himself, and he got it.

When he called CHNS radio as he'd been told to do, the call was immediately patched into the studio and Hayes's voice went on air live. He had no time to be frightened. With tears streaming down his face, Hayes broke the news to the world that rescuers had found a dozen miners alive deep within the bowels of the No. 2 mine at Springhill. That was the young reporter's moment in the global media spotlight. "Although I'd broken the story on both radio and newspaper, I was again reduced to a junior role as my seniors took over what had now blossomed into an even bigger story than it had been before," he said.⁴

It wasn't long before the other journalists in town learned the details of what was happening. Nor was it long until several senior *Chronicle-Herald* reporters rushed back to Springhill. Mere minutes after Hayes's broadcast on CHNS radio, Jack MacAndrew was heard on CBC across Canada delivering a news bulletin about what the media were already calling "the miracle at Springhill."

CHAPTER 24

"LAST MAN OUT"

—

WEDNESDAY, OCTOBER 29, 1958
4:00 P.M.

ARNOLD BURDEN WAS AT HIS McFARLANE STREET HOME,
enjoying what for him these hectic days counted as downtime. On
this glum, grey Wednesday afternoon, the doctor was lounging
around in a bathrobe and slippers as he sat filling out death certifi-
cates for some of the miners whose bodies he'd recently examined
and identified. It was depressing work, but it had to be done.

Burden had just written "multiple severe contusions and crushing" on yet another death certificate when the telephone rang. The
caller was one of the DOSCO managers. The man had startling
news to share: some men had been found alive in the mine.

"Can you come over to the mine immediately?" the caller
asked Burden.

"Yes, I'll be right there," the doctor said.

Within minutes, Burden was out the door and on his way to the
DOSCO offices. He arrived to find mine manager George Calder
seated behind his desk and talking on the telephone. Gesturing for
Burden to come close, Calder pointed to the piece of paper on which

he'd jotted down the names of the twelve survivors who'd been found alive. Among them was Caleb Rushton, who'd been married to Burden's sister and was still the doctor's friend. Beside the names of Joe McDonald and Ted Michniak, Calder had scribbled "severely injured." Seeing that notation underscored in Burden's mind the seriousness of the challenges he'd face if he went below and was able to care for the men. George Calder explained to him that so far, the only contact the rescuers had with the trapped men was through a length of broken air pipe. Burden already knew that—others had told him as much—but it didn't matter. He also knew that after almost six days in nightmarish conditions underground, the Twelve would be desperately hungry, dehydrated, and most likely in dire need of whatever medical aid he could render if he was able to reach them. Gorley Kempt had made clear how grim the men's condition was when he yelled through the pipe, "We need water. If we don't get some soon, you may as well stop digging and go home. We'll all be dead."

Burden advised Calder that he was ready to go down into the mine as soon as possible. That was just what Calder had been hoping to hear, and so it was with the manager's blessing that Burden headed for the pithead. By now, it was coming up to suppertime, and it was almost dark. The off-and-on cold drizzle that had made life in Springhill miserable all week was falling again. But Burden barely noticed as he hustled over to the mine, lugging his doctor's bag with him. He stopped first at the wash house to don overalls and then visited the lamp cabin, where he picked up a protective helmet and headlamp. Suitably attired, he strode along the muddy, oil-spattered roadway that led to the hoist shed. Many of the people who were clustered around the entrance recognized Dr. Burden; he was a familiar and much-admired figure in town. His mere presence was reassuring for it seemed to confirm there were men alive down below. Although they might be injured, there was at least some hope that not all the missing miners would be returning to the surface in coffins. Too many other men had already done that. And so people cheered and clapped when Burden passed. He responded with a wave before disappearing into the hoist shed.

Among the pieces of equipment that Burden had requested and that he took with him when he went down into the mine were 50 feet of copper tubing and a plastic hose. To this point, the rescuers had no idea how much farther they'd need to tunnel before they'd reach the men. Burden helped them figure that out. When he pushed the copper tubing through the 6-inch air pipe, he was surprised to find it was too short to reach the trapped men. A 100-foot length of tubing did the trick. After it was cut to the correct length, there was a 17-foot-long chunk left over. That meant the rescuers still had to dig through not 60 more feet of coal, rock, and debris before they'd reach the space where the Twelve were entombed, but rather 83 feet.

A 5-gallon pressure pump of the kind used to spray shrubs and trees was rushed to the mine from Amherst, solving the problem of how to get precious liquids to the men who so desperately needed them.

DOSCO's medical director, seventy-four-year-old Dr. J.G.B. Lynch, had advised George Calder to provide the men with a solution of water and vitamin B. However, when Arnold Burden heard about it, he disagreed and told Calder as much; vitamin B has a bitter, unpleasant flavour. Burden feared the men wouldn't be able to tolerate it on their empty stomachs and that it might even do them more harm than good. ("Have you ever bitten into a vitamin pill?" he asked.) Far better, said Burden, was to provide the men with plain water. Calder knew Arnold Burden well and trusted him without question; he opted to follow his sage advice. At 6:00 p.m., just four hours after Blair Phillips had made contact with the dozen trapped men, the first burst of water, cold and clear, began flowing through the copper tubing. However, to the dismay of one and all, when it did so, the effect wasn't as anticipated.

The water gushed through the line in an unexpected stream that shot out and spilled onto the mine floor. The dust there soaked it up instantly. It was like a cruel joke. It was impossible in the darkness for the men who were so desperately thirsty, weak, and disoriented to see or know exactly what they were doing. They had been blindly trying to fill their water cans. As a result, tempers were short. When Harold Brine put his mouth to the copper tubing and found it was dry, he was emotionally devastated; he lay down and wept. Gorley Kempt was so thirsty, he tried to lick the moisture off the mine floor. Several of the other men who got no water cussed Brine and Kempt. Fortunately, tensions were quickly defused when Burden called down the pipe to ask if they'd gotten the water. "No! We're in the dark in here. We couldn't see it," someone shouted.

"For God's sake, send us more!" another voice bellowed.

"Okay. Don't worry," Burden assured them. "We've got lots of water. We'll send you some as soon as you tell us when you're ready."

This time, they gathered water cans, found the end of the copper tubing, and were ready and waiting to capture the precious liquid when it came surging through the line. Caleb Rushton clutched his water can and muttered grace. "O, Lord," he said, "we thank you for the pipe and the blessed water."[1] Then, ignoring or having forgotten in the excitement of the moment Dr. Burden's caution to take only small sips and count to five hundred between them, like giddy children the thirsty men gulped great mouthfuls of water. "I came up with that number because I knew they would count fast," Burden would later quip.[2] A few minutes later, he sent sweetened black coffee down the copper tubing, and a while after that, he asked if the men would like some soup.

"What kind is it?" asked Gorley Kempt. Always quick with a quip, it hadn't taken long for his cheeky sense of humour to return.

"This is no restaurant!" said Dr. Burden, laughing. "It's tomato."

Revived by the fluids, the hot soup, and the realization that if only they could hold on for a few more hours, they'd be out of there, the Twelve waited patiently—or at least as patiently as they could in such trying circumstances. The rescuers still had a long way to go before they'd reach them. For the men who were trapped, every minute they waited to be rescued seemed like a day. Things were different with the barefaced rescuers. Prior to making contact, the men doing the digging had been advancing through about 12 feet of rock, coal, and soil in each eight-hour shift. "Once we knew there were live men on the other side, it only took [twelve] hours to go the remaining eighty-three feet," said Burden.[3]

The last few feet of tunnelling were among the most difficult. Harold Brine and Gorley Kempt repeatedly called down the pipe to ask how much longer it would be before the rescuers reached them. Harold Gordon, who continued leading the rescue efforts even though he was near exhaustion and suffering with a bad cold, told Brine he couldn't be sure how much longer it would be. The men doing the tunnelling had encountered a last-second obstacle: some rocks and timber from collapsed packs. They weren't sure what way to dig to get around the obstruction—above or below. It was Levi Milley, a veteran of more years in the mines than he cared to remember, who advised Calder that experience told him the rescuers had only another couple of feet to go; if they followed the broken air pipe by digging straight ahead, on a horizontal plane, they'd soon reach the chamber where twelve desperate men were waiting to be rescued.

DEEP UNDERGROUND, the rescuers were re-energized and working with a renewed sense of purpose to reach the dozen men trapped at the bottom of the 13,000-foot coal face. Three-quarters of a mile above, the air of anticipation was almost palpable. Police chief Leo

MacDonald and his two constables were busy making sure Main Street was clear of traffic and parked vehicles so ambulances would have an unobstructed path to rush injured miners to All Saints' Cottage Hospital, where a dozen beds were hastily being readied. At the same time, MacDonald and his men were assisting RCMP officers with measures to control the throng of anxious onlookers gathering at the pithead. Among the reporters were CBC television newsmen Lloyd MacInnis and Jack MacAndrew, who were live on the national network, reporting news of the latest developments. "They're talking miracles here in Springhill tonight," MacInnis told Canadians.[4] Indeed, they were "talking miracles" in Springhill, and the backdrop seemed appropriate to do so.

It was dark now, and the harsh glares of the floodlights needed for a nighttime outdoor television broadcast cast long, deep shadows across the landscape. The surreal scene on the surface was in sharp contrast to what was happening far below, where the drama that was unfolding was stunningly real.

"At 2:25 a.m. on that Thursday, October 30, we broke through. 'We're through!' resounded through the level. A blast of air and dust filled the tunnel, and after some minutes we could see," Burden would recall.[5]

As the rescuers clambered through the opening, their headlamps lit the chamber where the Twelve had been entombed. Several of them had risen to meet their saviours; Gorley Kempt did so, and to the surprise of one and all, he immediately scrambled out through the opening the rescuers had created. Those miners who were too weak to stand and the two who were seriously injured were lying on the floor of the mine. Dr. Burden rushed to Joe McDonald and Ted Michniak, while Burden's medical colleagues Ralston Ryan and Kenneth Gass, who'd come from Pugwash to join in the rescue efforts, tended to the other men. After quickly assessing their conditions, the doctors placed blindfolds over the eyes of all the men and loaded them onto stretchers.

When Gorley Kempt protested and asked why he needed a blindfold, Dr. Ryan told him it was to prevent any injury that might result from sudden exposure to the floodlights at the pithead. Kempt tore off his blindfold. "We've been down here in the pitch black for the last week; I don't want to be in the dark when I get up to the surface," he said.

It was almost an hour after rescuers reached the Twelve, at 3:23 a.m. on the Thursday morning, that one of the rescue crew, coal-blackened and exhausted but jubilant, emerged from the door of the hoist shed and waved for an ambulance to come close. He couldn't resist the urge to shout, "Gorley Kempt is on the way up!"

A buzz swept through the crowd, and then there were cheers as two rescue workers appeared while carrying Kempt's stretcher. A crush of people pushed forward to get a better look. "Kempt lifted himself to one elbow on the stretcher and waved, his teeth flashing white through a face as black as the rain-soaked ground. Then Kempt slumped back, exhausted by his show of bravado."[6] Moments later, the men carefully loaded Gorley Kempt's stretcher into the back of the ambulance, and he was rushed to hospital.

One by one, the other eleven rescued men were carried from the mine. All of them were as black as coal and bearded after almost a week without shaving. All were borne on stretchers, and all received roars of applause from the jubilant onlookers. Despite receiving warnings about keeping their eyes covered when they reached the floodlit surface, many of those rescued men felt themselves being swept up in the excitement. They peeked to see what was happening. How could they resist?

"We were laughing and crying all at once as they brought us to the surface," Fred Hunter told a *Chronicle-Herald* reporter. "They put a blanket over my face, but I showed them I wasn't finished. I took the blanket off and waved to the crowd."[7]

As the trolley car carrying Larry Leadbetter crested the top of the main slope and rolled into the hoist shed, he asked one of the men carrying his stretcher why there was such a lot of noise outside. Told there was a huge crowd of people and a slew of reporters waiting for him and the other rescued miners, Leadbetter laughed as he breathed a faux sigh of relief. "Phew," he said. "I was worried it might be bill collectors waiting for us."[8]

One of the last survivors to emerge from the mine was Harold Brine. He'd held onto George Hodges's hand during the entire forty-minute trolley ride up from the depths, and he did not let go until he was loaded into the ambulance.

When the last of the Twelve had been removed from their would-be tomb, Dr. Burden paused in that hot, smelly space. Even though there'd been some ventilation, the air was heavy with the musky scent of unwashed bodies, the putrid smell of human waste, and the sickly sweet stench of decaying flesh. In the next few days, searchers would recover the bodies of twenty-two more dead miners who'd met their end while working on this part of the coal face. Standing there alone in this now empty chamber, Burden listened intently for several minutes. One of the barefaced rescuers had reported hearing indistinct noises coming from the general direction of the top of the 13,000-foot coal face. Ted Michniak had also heard them. However, when Burden listened, he heard nothing. "At 4:20 a.m. I scribbled 'Last Man Out" [in my notebook] and with my aching muscles, [I] part walked and part crawled out of the black hole."[9]

CHAPTER 25

A PRINCELY VISIT

Thursday, October 30, 1958
5:02 A.M.

Following the rescue of the Twelve, the eyes of Springhill residents—and indeed those of the world—turned to the human dramas playing out at All Saints' Cottage Hospital.

Many of the townspeople who'd been at the pithead as the men were loaded into ambulances now rushed to the hospital where the traumatized miners would receive medical care. No sooner had the men shed their filthy garments and washed off coal dust and the smell of death before settling into their beds than their wives, children, and kinfolk flooded onto the ward. Close on their heels were newspaper reporters eager to interview the miners and their loved ones about "the miracle at Springhill."

Newspaper editorial pundits lauded the work of the fearless rescue workers—the draegermen and the barefaced miners—who were toiling tirelessly to save lives. "The survival and rescue of twelve Springhill coal miners is a reminder of the magnificent heights which the human spirit can scale," a *Globe and Mail* writer opined. "It is a reminder, also, of the great human will to [survive]."[1]

At the same time, a *Halifax Chronicle-Herald* editorial applauded "the unwritten code of the deeps to never give up hope"[2] for the miraculous rescue. When Canadians—and Americans too—read such paeans of praise, many of them were so moved that they reached for their chequebooks; hundreds of thousands of dollars in private donations poured in to the Springhill Disaster Relief Fund, and the federal government in Ottawa contributed $100,000. The money would be a godsend going forward.[3]

In the short term, however, most of the rescued men had fared better healthwise than most people had expected; miners as a breed are tough and resilient. To be sure, all the rescued men had lost weight during their ordeal—a minimum of 10 pounds. However, in hospital most of them soon regained that weight and recovered their vitality. That was especially true for the younger men who hadn't suffered any serious injuries. Larry Leadbetter had insisted he didn't need or want to be carried out on a stretcher. But it was a different story when he tried to walk; he staggered and nearly fell. Even Leadbetter acceded to being carried out of the mine and transported to hospital by ambulance.

Harold Brine was in a similar state. Twenty-six and fit as a prizefighter, he could walk unassisted, yet even he remained in bed for a couple of days while he regained his strength. Photographers delighted in snapping images of Brine as his daughter, Bonnie, sat on his lap while his wife Joan and sister Geraldine looked on.

Several of the older men weren't nearly as robust or fortunate as Leadbetter and Brine; their health had suffered more during their six-and-a-half-day ordeal, and it showed. These older men also found it took a longer time and a greater effort to recover. Ed Lowther was in this group. Lowther figured he'd seen it all in his years as a miner, yet he was an emotional wreck. Lowther couldn't stop crying. He had no idea why, but that's how it was for several weeks.

Joe McDonald, whose leg had been so badly broken, suffered

from a very different kind of emotional distress. He couldn't shake the notion that the men who'd been trapped with him—everyone except for his pal Ted Michniak—had cheated him when water was being doled out. "They thought we was going to die, and they just said, 'The hell with them,' and they didn't care whether we lived or died,'" said McDonald. "Boy, they're dirty, I'm telling you." He had a long memory. McDonald never forgave nor did he forget; he carried the grudge with him until it became a millstone around his neck. Many years later, McDonald would tell an interviewer, "When a person does a thing like that to you, then you ain't got no use for them."[4]

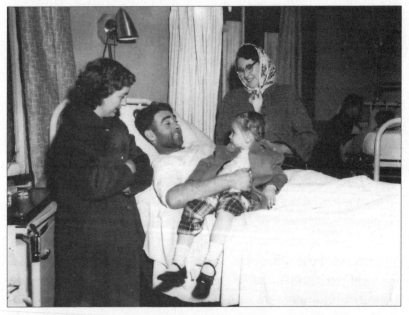

Harold Brine's wife Joan (*left*), daughter, Bonnie, and sister Geraldine visited him in hospital. (Clara Thomas Archives and Special Collections, York University)

Newspapers across North America featured lengthy stories about the twelve Springhill miners, as well as front-page photos of the men in their hospital beds. The banner headlines "First Up" and

"12 Saved" dominated page one of the *Halifax Chronicle-Herald*; "Miners Rescued" read the headline in the *Globe and Mail*, while the *Boston Globe* announced, "12 Saved—Will Live." The *Toronto Star*, ever Toronto-centric, featured a large photo of Harold Brine in his hospital bed along with a story that was rather badly headlined "Seek job for husband [*sic*] mine survivor's wife has eye on Toronto." Joan Brine's mother lived in Toronto, and so in the wake of the Bump, Joan was eager for Harold to find work there—anything that would enable him to leave behind his days as a coal miner.

The wives of the other rescued miners, older and younger alike, also spoke about how their husbands intended to look for new jobs. Hughie Guthro's wife, Margaret, expressed a common message when she said, "I don't care if we starve to death, Hughie will never set foot in another mine. There must be better things to do."[5]

Some miners made similar comments. "I'll never go into the pits again. I might take a correspondence course in auto mechanics or join one of the services," said Larry Leadbetter.[6] And then there was Levi Milley, who told a *Globe and Mail* interviewer, "If you're a newspaper fellah, you just put in your story that Levi Milley, former coalminer of Springhill and chicken raiser on the side, is looking for a full-time job on a chicken farm some place. Don't matter where."[7]

The sentiments expressed by those sadder but wiser miners and their spouses were by no means universal. There were still plenty of men in Springhill who were ready to return to work the day the No. 2 colliery reopened. That was understandable, for without the mine they had no jobs. John Scott, a forty-four-year-old veteran of the mines who'd survived three bumps, made that point in an interview with a newspaper reporter. Scott, who'd been badly injured on October 23 but had gotten out of the mine within hours of the Bump, spoke from his bed in the Halifax hospital to which he'd been transferred from Springhill. "Sure, I'll go back into the pits again," he said.[8]

Harold Cummings, another long-time miner, said it even more eloquently when he told the same journalist, "It's the only job we know. It's in our blood. It's our living."[9]

Despite the willingness of Scott, Cummings, and scores of other like-minded men to return to the mine, many people in Springhill were beginning to wonder if ol' No. 2 would ever reopen. They had ample reason to think that way. There were persistent, credible rumours that DOSCO's parent company, A.V. Roe Canada, was intent on closing the money-losing mine. On October 29, the Canadian Press news service had reported that Roe's senior managers had already made the decision to do so. Company chairman Crawford Gordon was quick to deny that news, but he did acknowledge the idea was on the table at the corporate headquarters in Toronto.

ALL THE TALK about the possible closure of the Springhill coal mine and about the fading hopes that rescuers would find any more men alive in the depths was temporarily forgotten when a royal visitor came to town. Prince Philip had chanced to be in Ottawa when the Bump happened, and on his return flight to London, at the Queen's request, he made a quick goodwill visit to Springhill.

The Comet airliner carrying the prince landed at Moncton airport on the afternoon of October 31. Then, accompanied by Nova Scotia Lieutenant Governor Edward C. Plow and a small entourage of federal and provincial politicians, the prince travelled the 60 miles to Springhill by car. By the time he arrived, it was about six o'clock and dark on that subdued Halloween night. "There were no cheering, flag-waving crowds for this royal visit, but there were 200 men, women, and children pressed close to the Duke as he passed through the pithead area [at the No. 2 mine]," the *Chronicle-Herald* reported.[10]

After a visit to the pithead and a short meeting at the mine with rescue workers who briefed him on their ongoing search efforts, Philip drove to All Saints's Cottage Hospital. There, Springhill mayor Ralph Gilroy, his wife, Grace, DOSCO manager Harold Gordon, and Dr. Carson Murray were waiting to greet him. Because the prince's visit was unplanned, there were no paparazzi or prissy British reporters watching as Philip toured the ward and stopped by the bed of each of the twelve rescued men. As a result, his ninety-minute visit to the hospital was low key and more informal than such royal events usually were. Reporter Val Sears of the *Toronto Telegram* described it as "what surely must have been the most informal reception ever given a Royal visitor in Canada."[11]

The first bed the prince came to was occupied by Harold Brine. When Philip leaned in to say hello, he quipped, "You don't look like you should be in bed. I hear you're the man who dug out one of his comrades." Brine wasn't sure how to reply when conversing with royalty, and so he just smiled and nodded. Philip chuckled. He then patted Brine on the shoulder and moved along to the next bed.[12]

The prince was relaxed and in good spirits as he made the rounds, talking to and joking with the injured miners, their family members, and hospital staff. Philip even ignored protocol when he signed some autographs. That was something members of the Royal Family seldom did—or do even today—owing to fears their signatures might be copied or forged. However, when Levi Milley, who knew nothing of that stricture, asked Philip for an autograph for his teenage daughter, the prince complied. "I don't do this as a rule," he said, "but I guess there's no harm in doing it this once." With that, he signed a piece of paper that he handed to Milley. Philip ignored the rules a second time during a subsequent quick stop at the Springhill Armouries, which were being used as an ersatz hospital. The prince signed the leg cast of George Hayden, one of the miners injured in the Bump.

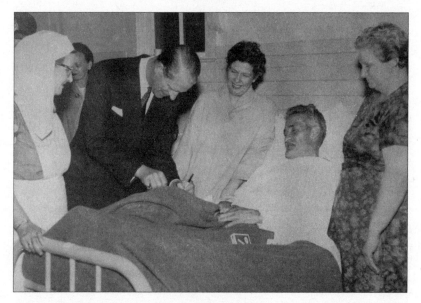

Prince Philip ignored royal protocols when he signed autographs for some of the injured miners he visited.
(Clara Thomas Archives and Special Collections, York University)

Before leaving Springhill, Philip made one final stop: a quick visit to the home of Mary Ann Raper, the grieving widow of a British-born miner. Fifty-five-year-old Harold Raper had been buried that afternoon. Afterward, Mrs. Raper, her children, and several neighbours gathered in the Rapers' living room to watch television coverage of the prince's visit to Springhill. "I was looking at him on television, and all of a sudden there he was in front of me," said Mrs. Raper.[13]

After they'd chatted for ten minutes, she offered to make the prince a "good cup of English tea." Philip politely declined. "Thank you, I'd have liked it," he said, "but I'm afraid I must run now." And with that he took his leave. Despite the royal's quick exit, Mary Ann Raper was honoured and overjoyed that he'd come calling. "He wasn't like a royal prince, but a prince of a man," she would later say.[14]

PRINCE PHILIP WASN'T the only well-known person to take an interest in the latest news out of Springhill. Those live radio and television news reports from the pithead and the extensive newspaper coverage spread awareness of the disaster across North America. As it happened, 2,500 miles west of Springhill, an American politician who was moose hunting as a guest of the Saskatchewan government chanced to see some of the CBC television reports. Georgia's Democratic governor Marvin Griffin was one of the most outspoken opponents of US President Dwight D. Eisenhower's efforts to promote civil rights and to intervene in what southerners euphemistically referred to as "the Negro question"—matters related to the Jim Crow state and local laws that enforced racial segregation in the southern United States.

Like most American governors, Griffin had no interest in events beyond his state's borders. However, he was always eager to boost tourism to Georgia and, of course, to do whatever he could to raise his own profile. Governor Griffin's ambitious young public relations (PR) man and chief speech writer knew this, and when Sam Caldwell heard about the mine disaster in Springhill, he came up with a promotional idea he figured was brilliant and would please his boss. As Caldwell saw it, his idea was so inspired that he was sure the governor of Florida or some other sunny state would hit upon the same idea and run with it. Unwilling to take that risk, without even running it past his boss, Caldwell dashed off a telegram to Springhill mayor Ralph Gilroy on behalf of "the Great State of Georgia." Caldwell invited the survivors of the mine disaster and their families to be Governor Griffin's guests in an all-expenses-paid recuperative week on Jekyll Island, an Atlantic coast resort community that Georgia was promoting. How could anyone turn down such a generous invitation? They couldn't.

After checking to make sure the offer was genuine, Mayor Ralph Gilroy acted on the miners' behalf when he contacted Caldwell to accept it. When word of the Georgia governor's generosity got out, it made the front page in newspapers far and wide. Caldwell was correct in thinking the trip would be a great promotional vehicle for his state. However, the moment the offer was extended, problems started to crop up. As the Scottish bard Robert Burns famously observed, "The best laid schemes o' mice an' men. . . ." What at first blush seemed like a surefire boon for the state of Georgia and its governor would soon go embarrassingly awry and become a public relations debacle for Marvin Griffin.

Meanwhile, as this drama was unfolding in Georgia, in downtown Toronto A.W.J. Buckland, the editor of the *Toronto Telegram*, was watching events in Springhill with keen interest. "The Tely"—as it was known on the street—was the self-proclaimed champion of working-class, conservative, Orange Toronto, and as such it was the bitter rival of the liberal *Toronto Daily Star*. The two papers waged a spirited circulation war. Both newspapers sent reporters to Springhill; however, as public interest in the dramatic search for missing miners grew, the *Telegram* increased the scope and volume of its coverage. The articles garnered prominent placement in each day's edition, and they won wide readership. In some measure, this popularity was due to the extensive radio and television coverage of the disaster, which heightened people's interest in the dramatic events happening 1,000 miles to the east. And in some measure, reader interest in what was happening in Springhill was stoked by the tone of the *Telegram*'s coverage, which was often lurid. Typical was the November 1 banner headline on the story about the rescue of the Twelve: "The Diary of 12 Dead Men."

When that day's paper was published, it was still too soon to know if any more men would be found alive, but Buckland was already looking ahead. He knew once the last of the Bump survivors was rescued, the disaster story would come to an end unless he could find some way to keep it alive. With that in mind, Buckland and his editing team devised a plan to get some of the rescued Springhill miners on national television. What better way to do that than to have them appear on a popular American network program? There was one that seemed ideal: *The Ed Sullivan Show*, the eponymous vaudeville-style variety show that aired live each Sunday night from 8:00 to 9:00 p.m. Eastern Time on the Columbia Broadcasting System. Most weeks, more than twelve million viewers in the United States tuned in, and the show was also phenomenally popular in Canada.

Apart from the curiosity factor involved in having rescued Springhill miners appear on that show, Buckland hoped the exposure would create awareness of the Springhill Disaster Relief Fund—and, of course, the newspaper's role in promoting it. There were several reasons the *Telegram* chief thought he could sell Sullivan on the idea of promoting the relief fund.

One was that the show's emcee was a fifty-seven-year-old former newspaper columnist with a blue-collar mentality. While Sullivan himself wasn't a performer, he was a star. On stage, he was stiff as a mannequin and spoke with a slight lisp; his signature catchphrase—about having a "really big shew tonight"—was part of the North American lexicon from the time the program debuted in 1948 until it ended its twenty-three-year run in 1971. Sullivan's occasional verbal gaffes on live television also endeared him to viewers. Because the show was live, there was no taking back or editing his words. Hence audiences enjoyed watching as things went sideways on air and Sullivan messed up, as he did the night he asked singer Connie Francis, "Tell me, Connie, is your mother still dead?"

Apart from the fun of watching a live television show in which anything could and did happen, what made *The Ed Sullivan Show* must-see viewing for millions of people was that each week Sullivan showcased rising stars everyone wanted to see, and sometimes their performances were historic—such as those by the Beatles, Elvis, Bob Dylan, a young Michael Jackson, the Rolling Stones, and many others.

Another reason for the popularity of Sullivan's show was his Everyman sensibilities. He may not have always understood why a performer had caught the public's fancy, but he was savvy enough to take advantage—that is, unless he took a dislike to the performer or felt disrespected by him or her for any reason. The show was Sullivan's signature initiative, and so he ran it on his terms. That approach nicely tied into the other compelling reason the editors of the *Toronto Telegram* thought Sullivan would be eager to have some Springhill miners as guests on his show.

It so happened on the Sunday night after the Bump, October 26, one of the performers on *The Ed Sullivan Show* had been Shecky Greene, a Las Vegas nightclub comedian whose shtick was wisecracking, storytelling, and improvisation. That night, just as Greene was about to step on stage to do his eight-minute stand-up routine, Sullivan let him know the show was running late; Greene's time slot had been cut to just two minutes. The comedian was taken aback and went on stage flustered, angry, and in full improv mode. Toward the end of his abbreviated appearance, he pretended to be talking through a tube to someone under the stage. Maybe Greene had heard about a disaster at a coal mine somewhere in Canada called Springhill.

"What's that? You say the canary died? That's a gas!" said Greene. "You say you're hurt? Nah, you weren't hurt. That was *soft* coal!"[15]

Ed Sullivan was waiting for Greene when he came off stage, and the emcee was angry. When he'd heard Greene joking about coal miners being gassed, Sullivan's mind had immediately gone to the tragic events in Springhill that he'd been reading and hearing about. Sullivan proceeded to upbraid the comedian, calling him "the sickest son of a bitch I've ever known in my life." Greene insisted he didn't know what he'd done wrong, but that didn't soothe Sullivan's anger. He was acutely aware of his audience and the ratings for his show. "You lost me Canada!" he snarled. "You'll never set foot on this stage again."[16]

Sullivan was true to his word, for he had a memory so long it would have made an elephant seem forgetful in comparison; Greene never appeared on *The Ed Sullivan Show* again. At the same time, Sullivan was keen to make amends with those Canadian viewers who'd taken offence at Greene's attempts at humour. There were many. When someone from the *Toronto Telegram* called New York to ask if Sullivan might be interested in having some Springhill miners exclusively on his show and in helping to promote their disaster relief fund, Sullivan was receptive. He'd be delighted to do both, he said, and he wanted some of the rescued miners to travel to New York, all expenses paid, for an appearance on his show that Sunday night, November 2.

One of the recurring features on *The Ed Sullivan Show* had Sullivan standing on the stage apron while inviting celebrities, newsmakers, and other public figures who'd been strategically seated in the audience to stand up, wave, and take a bow. That was something Sullivan envisioned Springhill miners doing, and in the process, he'd be making amends with Canadian viewers for Shecky Greene's tasteless jokes. It would be a win both for the miners and for Sullivan.

CHAPTER 26

SEVEN UP

SATURDAY, NOVEMBER 1, 1958
3:30 A.M.

THE EUPHORIA THAT CAME WITH THE MIRACLE RESCUE AND
Prince Philip's visit to town faded as quickly as a morning fog.
When it did, the people of Springhill again found themselves facing
the grim realities of the disaster that had befallen them and their
town.

In the hours after the rescue of the Twelve, barefaced miners
continued working to recover bodies from the mine; many men
were still unaccounted for. A small crowd of women and children who hadn't forgotten that—and who'd never let themselves
do so—maintained their vigil at the pithead. Although they were
wet, cold, and despondent, they refused to abandon their hope for
another "Springhill miracle." However, that now seemed less and
less likely, especially after the Friday morning television broadcast in which DOSCO vice-president and general manager Harold
Gordon delivered more deflating news.

He'd reported that although gas levels in the mine had abated
and the three crews of barefaced miners—about seventy men—

were working round the clock, they were making slow progress in their rescue and recovery efforts. Gordon feared it could be another week before they'd reach the top end of the coal face on the 13,000-foot level. That was where any men who were still missing most likely would be found. Alive or dead.

This post-Bump photo, taken by Herbert Zorychta of the Mines Branch of the federal Department of Mines and Technical Services, shows the damage to the tunnel that was at the 13,000-foot level of the No. 2 mine. (Courtesy of the Burden family)

Gordon had been in charge of the underground rescue mission from the start; DOSCO's chief mining engineer, Louis Frost, was in town, but he left in Gordon's capable hands the organization and details of the day-to-day efforts of the barefaced miners and draegermen. With the eyes of the world on Springhill, the responsibilities Gordon shouldered were onerous, and it took a heavy toll on him, both physically and mentally. At one point, DOSCO's medical chief, Dr. J.G.B. Lynch, had ordered Gordon to rest. He reluctantly did so, taking just one day off before he went down into the mine again. Gordon was there with the barefaced miners in the wee hours of Saturday morning.

It was about 3:30 a.m., and the crew was bone-weary after yet another long night of sweating, cussing, and digging to carve out a rescue tunnel. Deputy overman Bill "Bud" Henwood had been doing the spadework for a couple of hours, and he needed to rest. He found it murderously hard to lie on his stomach for hours on end while chipping away at the rocks, coal, and bits of pack timber that blocked the route to the coal face on the 13,000-foot level. When Henwood was backing out of the tunnel for a break, Harold Gordon, crew chief Bill Miller, and the other men who'd been on the bucket brigade repositioned themselves

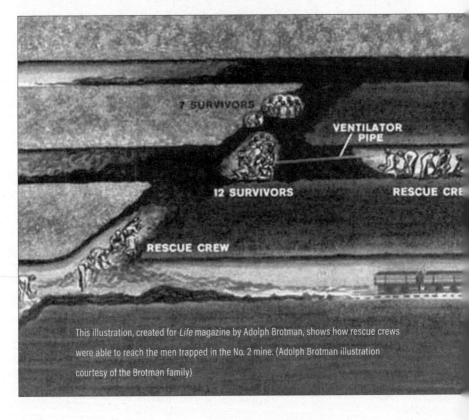

This illustration, created for *Life* magazine by Adolph Brotman, shows how rescue crews were able to reach the men trapped in the No. 2 mine. (Adolph Brotman illustration courtesy of the Brotman family)

to allow him to retreat. As they were doing so, there was an interval of relative quiet. It was then that there came a scratching sound. It seemed to be coming from somewhere ahead of where they'd been digging.

"I think I heard something," said Henwood. "Did anybody else hear it?"

A few other men nodded and said yes. "Rats. It's probably just rats," one of them replied.

However, the longer they listened, the less the noise sounded like rats doing what rats do. At one point, there were three slow, distinct intervals of the scratching.

"No. That's not rats. It sounds like somebody is digging or trying to send a signal," said Gordon.

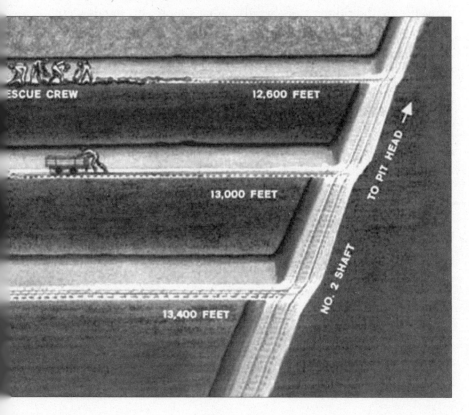

That was all the impetus Bud Henwood and the rest of the rescue crew needed to redouble their efforts. Henwood dived back into the tunnel, and for the next fifty minutes, he and the rescue crew dug like men possessed. Then, suddenly and without warning, when Henwood swung his pickaxe, it broke through into an open space. He frantically hacked away to make the opening bigger. When the beam from Henwood's headlamp stabbed into the darkness and the fog of coal dust that clouded the air in the chamber in front of him, he could scarcely believe what he saw.

There, lying flat on his back and wedged into a small crevice, was a miner with his legs jammed up under him like a contortionist. The man was moaning. His eyes were wide open, yet his gaze was unfocused. Henwood could see the injured miner was hanging onto life by the thinnest of threads. What's more, his hands were a bloody mess. All the skin was gone from his fingertips, and the nails had been worn down to the quick by his anguished efforts to scratch and claw at the rocks that imprisoned him.

"Oh my God! I've found somebody, and I think he's alive!" Henwood shouted. He could barely contain his excitement. "Send for Dr. Burden! Have him come as fast as he can."

ARNOLD BURDEN WAS at home when his telephone rang early on that Saturday morning. It was still dark outside, but the doctor was up and around. He was busy packing a suitcase for the trip to New York. He and his wife would be departing in about twenty-four hours. Arnie Patterson, DOSCO's public relations chief, had arranged for miners Gorley Kempt and Caleb Rushton to make the all-expenses-paid trip to appear on *The Ed Sullivan Show*; both men had recovered well from the ordeal of their six-and-a-half-day entombment. Patterson had also wrangled invitations for himself, Gorley's wife Margie, Caleb's wife Pat, and Dr. Burden and his wife Helen. Burden ostensibly was going along to care for Kempt and Rushton. Truth be told, Patterson viewed the trip as a small reward for Burden's selfless contributions to the relief efforts at the mine in 1956 and again this time around.

Burden forgot all about the New York trip the moment his caller asked him if he could come to the mine and provide emergency medical care to a miner who'd been found alive on the 13,000-foot level. Burden was "shocked down into my shoes," as he would later state. "The call came at 4:45 [a.m.] and at 5:00 I was at the pit. At 5:12 I went down on a trolley with a group of barefaced men."[1]

After scrambling along the rescue tunnel, Burden squeezed through the small opening that served as the entrance to the chamber where the injured man was lying belly up. A couple of rescue workers—John Calder and Bennie Roy—were still busy moving rocks and coal so they could pull the man out of the crevice into which he was wedged. "We think it's Barney Martin," Calder advised Burden.

It was impossible to verify that just now. The injured man was semi-comatose, black with coal dust, and he was in no condition to talk. Understandably so, for he was badly dehydrated and barely breathing. Burden helped Calder and Roy load him onto a stretcher, and after flushing out his mouth with a splash of coffee, they rushed him down the tunnel and out of there. Any assessment of the man's injuries would have to wait until he got to the hospital.

In the meantime, Burden, Calder, and Roy began picking their way through the debris that littered the coal face on the slope below them. In some places, the headroom was so low that the rescuers had to stoop in order to proceed. Their uncertainties and apprehensions mounted with each foot they advanced.

Ahead of them in the darkness, Maurice Ruddick and the other five men who were still alive—just barely so—were lying on the floor of the mine and waiting for death to take them. Deprived of adequate food and water, and with some of them in dire need of medical care, the men were moribund. They were as lifeless as the aura in the hellish chamber in which they'd been entombed for eight-and-a-half days. The stench in the air was stomach churning and fetid beyond measure. It reeked of human waste, unwashed bodies, and the sickly sweet, heavy smell of decaying flesh.

By now, Maurice Ruddick was oblivious to it all. He was dreaming of his home and family as he drifted in and out of consciousness. What was real and what wasn't? It had become difficult for him to tell. That's when he saw bobbing in the distance what appeared to be three miner's headlamps coming toward him in the darkness. He could scarcely believe his eyes. Although his legs felt rubbery, Ruddick rose to one knee. Then, after placing a hand on the mine wall to steady himself, he somehow found the strength to stand on the uneven floor. The next thing he knew, a blinding light was shining in his eyes, and a voice was greeting him. At almost that same moment, Garnet Clarke grabbed him by the head and kissed him on the cheek. "Thank the Lord! We're saved, Maurice! We're saved!" he said in a gravelly whisper.

"How many of you are in here?" asked John Calder.

"Six. There are six of us," someone called out in the darkness.

"Who needs medical help?" asked Arnold Burden.

Later, it would be reported that when the rescuers approached Maurice Ruddick, he'd said, "Give me some water, and I'll sing you a song." However, Ruddick would have no recollection of having said those words; they most likely were the product of a journalist's imagination. About all Ruddick could confirm was that during the eight-and-a-half days he and his comrades had been trapped, they'd sung more hymns than they could count. "The Old Rugged Cross" and "Rock of Ages" were two of the songs Ruddick had found to be particularly comforting.

When Dr. Burden passed him a water can, Ruddick grabbed it and took a long, deep gulp. Meanwhile, Calder and Roy had offered water cans to the other men. Currie Smith, Herb Pepperdine, Doug Jewkes, and Frank Hunter were like thirsty camels. After so many days, the men had never tasted water so delicious That was so even for Doug Jewkes, who still had a craving for 7Up.

The men's thirsts having been quenched, at least temporarily, John Calder turned to Ruddick, who hadn't yet burst into song. Arnold Burden laughed as he heard Calder advise Ruddick that the provincial Workmen's Compensation Board had sent him into the mine with a specific mission: find Maurice Ruddick, and get him out of there alive.

"What? Why did they do that?" Ruddick asked.

"They said if we don't find you, they'll have to pay so much for your wife and 12 kids that there won't be enough for the other men and their families," Calder replied with a grin.[2]

For the first time in several days, Maurice Ruddick and his comrades had a reason to laugh. They were still doing so when other members of the rescue team arrived. "As soon as our lights entered the area, one of the trapped men became extremely agitated. It was the sight of his dead buddy. He said, 'If that body isn't covered, I'm going to go out of my mind,'" Burden would report.[3]

The decaying body of Percy Rector, who had died four days earlier, was still hanging from that collapsed pack. The beams of the rescuers' headlamps illuminated with staggering clarity the horrors of Rector's suffering and death. There is no dignity in death, and even less when it's the result of the kind of anguish Percy Rector had endured. Both his suffering and his passing had been nightmarish beyond measure. Recovering his corpse would be a grisly task for the barefaced miners who'd vowed to leave no man behind in the No. 2 mine, dead or alive. As always, these heroes were true to their word.

However, for now the recovery of Rector's decomposing body had to wait. The priority for Dr. Burden and the rescue team was saving the six men who'd been found alive. After providing each of them with water, a serving of hot chocolate, and then some tomato soup, rescuers helped them over to the opening of the access tunnel through which they then scrambled to reach the 13,000-foot level. There other rescuers loaded them onto stretchers for the trolley ride up to the pithead and a reprise of the scenes of jubilation there and at All Saints' Cottage Hospital that had greeted the rescue of the Twelve two days earlier.

Word of the second "miracle at Springhill" came as a shock to many townspeople. In fact, it was so unexpected and it happened so quickly that some people—even family members of the rescued men—didn't hear the news for an hour or more. In at least one case, they heard the news indirectly. When Wally Hayes, the young *Halifax Chronicle-Herald* reporter-photographer, called at the home of one of the miners to get a comment from the man's wife, Hayes got almost as big a surprise as the woman did. "When I arrived at the house, there was a crowd of people listening to a radio news report about the rescue," said Hayes. "I asked the lady-of-the-house how she'd felt when she'd heard that her husband was among the survivors who'd just been rescued. She nearly fainted. She hadn't heard the news, and so she grabbed me and gave me a big hug. She cried, and so did I."[4]

Once the names of the seven men who'd been rescued became known around town, their loved ones and friends rushed to the hospital, and so too did the few journalists who had remained in Springhill. Several of the rescued men garnered special attention from the media; Barney Martin was prominent among them. Martin recounted his experiences in an interview with a *Toronto Star* reporter who crafted a lengthy feature article. Then there was Doug Jewkes. When news of his craving for 7Up reached the ears of the soft drink's local distributor, who was in Moncton, he rushed a carton of the drink to All Saints' Cottage Hospital. Jewkes was delighted to pose with bottles of the beverage when newspaper photographers asked him to do so. And he was even happier when, a few days later, he received the offer of seventy-five dollars a week working in the bottling plant of the Toronto distributor of 7Up.

Doug Jewkes enjoyed a complimentary case of 7Up that was given to him by the soft drink's local distributor while he was in hospital following his rescue. (Clara Thomas Archives and Special Collections, York University)

There was no shortage of people and companies eager to bask in the reflected glow of the public interest in this latest "miracle at Springhill." For several reasons, no one received a larger share of that media attention than did Maurice Ruddick. One was because his four-year-old son, Revere, chanced also to be a patient at All Saints' Cottage Hospital; the boy had been admitted after coming down with red measles. When a *Toronto Telegram* newspaper photographer heard this, he asked Ruddick to pose with his son, and they did. The photo of the two of them, cheek to cheek and staring at the camera, appeared the next day in the *Telegram* and scores of papers across Canada and the United States.

Ruddick also became a focal point of media attention in no small measure because of a comment made by Arnie Patterson, DOSCO's communications director. While talking about how and why the Seven had survived for more than eight days in such ghastly conditions, he explained, "Maurice was the leader of the group. [He] was the top miner on the level, and he was the chief joker," said Patterson. "He kept [the men] laughing, and he kept them singing." Doug Jewkes echoed that message when he told a reporter, "Maurice Ruddick . . . he's the one that kept our hopes up and us alive."[5]

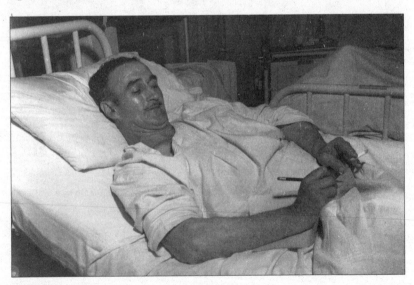

Maurice Ruddick used his time in hospital to write a song about his post-Bump experiences.
(Nova Scotia Archives)

While Patterson and Jewkes were making those pronouncements, Ruddick—whose skin colour and bearing also set him apart
in the eyes of those journalists who were looking for "an angle,"
real or invented—was lying in his hospital bed, cigarette in one
hand, a pencil in the other as he jotted down the lyrics to the song
about the Bump that he'd been mentally composing for the last
week. "There's a hard luck mining town in Nova Scotia / Whose
story is sad with fame . . ."[6] It was inevitable that reporters would
seize upon that and tie it to a description of Maurice Ruddick as
"the Singing Miner." And so it was that the name Ruddick had
adopted when he was performing on stage became synonymous
ever after with his surname. A legend was born.

WITH THE SEVEN now resting in hospital, Dr. Burden returned
home after fifteen gruelling hours down in the mine. He was tired,
and since it was now about ten o'clock on the Saturday night, the
doctor and his wife, Helen, had little time for sleep. The Burdens,
Arnie Patterson, and the two miners who were to appear on Sunday
night's *Ed Sullivan Show*—Gorley Kempt and Caleb Rushton—
had to be up and ready to leave for Moncton airport by 5:00 a.m.
to catch the first leg of their connecting flight to New York. On
board the plane, they received celebrity treatment. The invitation
from the Trans-Canada Air Lines pilot to visit the cockpit was an
added thrill to what was already a memorable experience. This
was the first time the Kempts and Rushtons had flown, and it was
the first time they and the Burdens had been to New York City.
Adding to Dr. Burden's sense of adventure was the fact he had
almost no money in his pocket and only a couple of pictures left on
the camera he was taking with him to record his trip.

Remember, in 1958 most stores were closed on Sundays (and
Wednesday afternoons in many parts of Canada). There were no

bank machines, and the institutions themselves operated on what Canadians often derisively referred to as "bankers' hours"—10:00 a.m. to 3:00 p.m., five days a week. What's more, few people had credit cards in 1958; it was in that year that American Express introduced the first plastic credit cards for widespread consumer use in Canada and the United States.

Fortunately for Burden, a big-city newspaper reporter who chanced to be aboard the same plane out of Moncton gave him a couple of rolls of film. However, that did nothing to lessen the chagrin Burden felt to be travelling with no money. His wallet was all but empty of everything other than air when he met the millionaire Canadian newspaper magnate Max Aitken, better known as Lord Beaverbrook, on the flight from Saint John to Boston. Beaverbrook had heard about events at Springhill, and he was eager to meet Arnold Burden and the two miners he'd helped rescue. "A wonderful thing," said Beaverbrook. Then, being the quintessential newspaper publisher, he hastened to add, "and a wonderful story."[7]

The Springhill delegation reached New York by early Sunday afternoon, and things went smoothly there. Ed Sullivan and his staff made sure the Canadian visitors were well cared for and made to feel welcome. They had rooms at the upscale Park Sheraton Hotel, in the heart of Manhattan, and they were provided with all the amenities they could have wished for; the Burdens took advantage of room service, dining on sirloin steak. A hotel chambermaid asked the Springhillers, "Are you the men God Almighty saved?"[8] And out on the streets of New York, when passersby and cab drivers recognized the Burdens, Kempts, and Rushtons, they were feted like the overnight celebrities they were; Sullivan's staff even saw to it that they were given tickets to see the hit Broadway show *My Fair Lady*.

The only negative during the two-day visit was some of the intrusive media attention, which included several late-night calls to their hotel rooms by nosy reporters and some approaches on the street by photographers for the various newspapers. "[They] dogged us during our entire visit," Burden would recall. "Some were just plain jerks, asking us to look up at tall buildings as if we were country hicks who had never seen buildings before. I told them to go to hell."⁹

As for the group's Sunday night appearance on the Sullivan show, it went off without a hitch. Partway through the hour, Sullivan stood on the apron of the stage holding up a copy of the previous day's edition of the *Toronto Telegram*—the one with a banner headline that read: "The Diary of 12 Dead Men." In a stroke of marketing genius, *Telegram* reporter Phyllis Griffiths, who'd joined the Springhill crew in New York, had supplied Sullivan with the newspaper.

After referencing the *Telegram* article, Sullivan pointed to the audience and introduced Dr. Arnold Burden, Gorley Kempt, and Caleb Rushton. He asked each of them in turn to stand up and take a bow. After they'd done so, Sullivan issued a direct appeal on behalf of the Springhill Disaster Relief Fund and provided the address where viewers could send donations. The entire segment took only slightly more than two minutes—about the same amount of time comedian Shecky Greene had been on stage the previous Sunday when he seemed to make light of events in Springhill—but Sullivan's appeal had far more impact.

Donations to the disaster fund poured in from across Canada and the United States, including more than $100,000 from A.V. Roe, DOSCO's corporate parent.¹⁰ The total amount raised would eventually top $2 million—about $21 million in today's currency. Ed Sullivan himself was among the donors. Backstage after the show, he handed Burden a personal cheque for $1,000; that was his contribution to the cause. Then Sullivan signed autographs for the Kempt, Rushton, and Burden kids, all of whom—just like almost every resident of Springhill—watched the show with rapt attention.

CHAPTER 27

BE CAREFUL
WHAT YOU WISH FOR

Thursday, November 6, 1958
10:00 A.M.

There was one question that kept coming up during the Springhillers' visit to New York City. Again and again, people approached them at the hotel and on the streets to shake hands and ask: "Are they going to find any more men alive down in the mine?" Sadly, the answer was "No. They won't. There won't be another Springhill miracle."

That sad reality was confirmed by the barefaced miners who'd continued to explore the depths of the shattered mine. They were finding no more men alive, only silence, darkness, and human remains. Harold Gordon confirmed as much when he announced that the last thirteen missing men had all been located, and all were dead. This brought the death toll in the Bump to seventy-four. It would rise by one more when an injured miner—fifty-seven-year-old Springhill native Bill Stevenson—died in hospital on November 22.

In the wake of so many deaths and the November 12 announcement that A.V. Roe Canada, DOSCO's parent corporation, had decided to close No. 2 mine because it was deemed no longer safe to operate, the people of Springhill felt devastated. After eighty-five years as a coal producer, their town had lost both its identity and its principal employer. Nine hundred jobs vanished overnight. "This looks like the knock-out blow [for our town]," said Mayor Ralph Gilroy.[1] Gilroy, local business leaders, and federal and provincial politicians vowed to do whatever it took to attract new industries and bring new jobs to Springhill, but it was an uphill, quixotic battle. Small wonder that an oppressive, all-consuming cloud of despair hung over the town.

The pages of the *Springhill-Parrsboro Record*, the town's weekly newspaper, were filled with column after column of obituaries of the Bump's victims. The local funeral parlours were the only businesses that were busy. For more than a month, not a day passed when residents didn't see funeral cortèges wending eastward along Main Street, past Miners Memorial Park, en route to Springhill's burial grounds. Hillside Cemetery had to be expanded to make room for the spate of burials.

Meanwhile, over at the pithead, the crowds of local women and children who'd kept up a vigil for the still-missing men had finally melted away, their hopes now forever dashed. In their place were sightseers from neighbouring towns who came to indulge their morbid curiosity about what was happening in Springhill. These gawkers watched the barefaced miners who came and went from the mine and who periodically emerged from the hoist shed to load yet another aluminum coffin onto a DOSCO truck. As the *Toronto Telegram* reported, "A weary miner stepping out of the pithead took one look at the laughing, chatting people dressed in their Sunday clothes and said, 'There's not a Springhiller among them.'"[2] That miner had it right.

Barefaced miners recovered the body of the final missing

man—Fidell Allen, Maurice Ruddick's forty-year-old neighbour on lower Herrett Road—from the coal face on the 13,800-foot level on the evening of Thursday, November 6. That was two long weeks after the Bump. It was also the day the last of the big-city journalists who'd been there to report on the disaster left town. Wally Hayes, the young *Halifax Chronicle-Herald* reporter-photographer, returned home sadder but wiser. "I'd learned a lot in my two weeks in Springhill. One of them was that I was becoming hardened, callous even, after being around so much death and suffering. That was an eye-opener for me," said Hayes. "And I didn't like the feeling."[3]

Sadie Allen prayed and waited in vain for the safe return of her miner husband, Fidell, whose body was the last to be recovered from the No. 2 mine. (Clara Thomas Archives and Special Collections, York University)

Like many people, Hayes felt worn out, drained by the emotional highs and lows he'd experienced. But he could leave the sense of loss and heartache behind; the locals didn't enjoy that same luxury. A weariness had set in among many of the people who'd endured the pain and profound sense of loss that came with the whole experience. Some of the nineteen miners who'd been rescued—along with Dr. Burden, who was their special guest—quietly got together at the home of Joe McDonald, the last man to leave hospital. Most local people were eager to forget what had happened and to move on with efforts to repair their broken lives. Those outsiders who'd followed events from a distance, reading the daily newspaper articles or tuning in to the radio and television news reports, were of a different mind. They remained enthralled.

There was still considerable curiosity about the Springhill mine disaster and stories of the survivors' experiences and prospects. Ed Sullivan had mentioned that the deaths of the Springhill miners had left 158 children fatherless. This prompted donors from across North America to reach out to some widows of men who'd perished; among them was Ruth Tabor with her four children. "I can remember that there was an older couple in Campbelltown, New Brunswick, who sent my mother money. They'd seen an article in the newspaper," recalled Valarie (Tabor) Alderson. "Every year for a lot of years, the couple sent a cheque at Christmastime. Mom always made something to send back to them. They even came to visit one time."[4]

As Ruth Tabor's story suggests, the tone and content of the media coverage of the Springhill mine disaster changed, as did the focus of the barefaced miners who continued to go into the mine each day. What had been a rescue mission was now a joyless operation to recover the dead. Journalists moved on from reporting the soul-sapping details of the disaster and the miraculous rescue of those nineteen miners. Instead, they turned their attention to the stories of the men who'd survived the Bump and of those who were now joining in the mass exodus from Springhill. Scores of them packed up and moved away in hopes of putting their sorrows behind them and starting new lives elsewhere. There was nothing left for them in Springhill. That's how Harold Brine and his wife Joan felt.

The mine's closure left Brine out of work, strapped for cash, and directionless. With few job prospects locally and no money to pay for the lumber and other materials he needed to finish the house he was building, at age twenty-six Brine was ready to abandon his life in Springhill. On November 3, just a few days after he got out of the hospital, Harold Brine, his wife, and their daughter made a flying visit to Toronto where Harold appeared at a $100-per-plate fundraiser for the Springhill Disaster Relief Fund. Predictably, the Brines' visit to the Big City attracted a lot of media attention.

When the Brines arrived in Toronto, they found newspaper reporters, photographers, and television interviewers waiting for them. This *Toronto Telegram* photo of the family appeared in newspapers across Canada the next day. (Clara Thomas Archives and Special Collections, York University)

When their Toronto-bound Trans-Canada Air Lines plane stopped to refuel at Montreal, CBC television broadcaster Joyce Davidson rushed onto the tarmac to interview Brine. This promoted an angry outcry from Toronto television newsman Harvey Kirck, who was waiting for the Brines at the Malton Airport in Toronto (now Pearson International Airport) for what he'd anticipated would be an "exclusive." Newspaper reporters and photographers from the Toronto newspapers were also there waiting. When the Canadian Press news service picked up the photo of the Brines that accompanied the *Telegram* news story, it appeared in newspapers across the country.

ANOTHER PECULIAR ASPECT of the post-Bump experiences of Springhill miners that generated headlines was that unexpected invitation from the state of Georgia for the men who'd been rescued—plus their wives and kids—to enjoy a week-long, all-expenses-paid vacation at Jekyll Island State Park. When the original offer to host the miners was made, only twelve men had been rescued. Behind the scenes, the governor was lukewarm to the initiative. He warmed to it only when he saw that just two miners had appeared on *The Ed Sullivan Show*. Griffin was buoyed by the thought that the other ten men might still be in hospital or not well enough to make the trip to Georgia. Then too, the governor saw how much public interest there was in the planned trip and how great the possibilities were to promote tourism to his state. However, as mentioned earlier, the governor's plans quickly began coming apart.

First, officials at the airline that had offered to fly the miners and their families from Nova Scotia to Georgia gratis reneged when they discovered that US aviation regulations forbade any airline from offering free flights. Then came the news that seven more

Springhill miners had been rescued. This brought the total number of invited guests to nineteen, plus their wives and children. Griffin was aghast when he realized the potential cost of flying dozens of people from Springhill to Georgia. In a bid to make the best of a worsening situation, he told a *Toronto Daily Star* reporter, "There's lots of room for everyone there [on Jekyll Island]."[5]

That was all well and good until Griffin learned one of the rescued miners, Maurice Ruddick, was "coloured." This gave the governor "the surprise of his life," as one journalist wrote. It also posed an embarrassing public relations problem for Georgia's governor.[6] The state has a long, troubled history of race relations; among America's fifty states, Georgia is second only to Mississippi in its record of lynchings of Black people—458 to 538. And Griffin himself was a staunch segregationist who'd vowed to close the state's public school system if the federal government tried to enforce desegregation.[7]

Griffin understood it would reflect badly on him and "the Peach State" if he withdrew his invitation to the Canadian miners. It might also do irreparable harm to Georgia's tourism sector. And so, in a bid to appear conciliatory while defending the undefendable, Griffin announced that his invitation would stand only if Ruddick, his wife, and their children agreed to stay in quarters separate from the white miners and their families.

When word of this ultimatum reached Springhill, the miners were divided in their reactions. Most were ready to stay home if Maurice Ruddick and his family were forced to stay in segregated accommodations, eat meals in a "coloureds only" dining area, and swim only at an isolated section of the beach, well away from the area whites were using. Harold Brine, for one, was adamant that if Ruddick refused to comply with those humiliating conditions and decided not to visit Georgia, he and some of the other miners wouldn't go either. "We said, if you won't accept him, we're not coming," Brine recalled.[8]

Said Bowman Maddison, "I want to go. If Mo says it's all right, we'll all go."

Herb Pepperdine had a similar reaction. "I still think if [Ruddick] is good enough to work with us, he'd be good enough to holiday with." Then he was quick to add, "I just can't make up my mind."[9]

However, several of the other miners were willing to accept the trip on the governor's terms; after all, for most of them it would be their first southern vacation, the farthest from home they'd ever been, and their first time to fly anywhere. The trip was a big deal. Said Ted Michniak, "Up here, we work together and we go to school together," he said. "But if there are different laws down in Georgia, I guess we should abide by them."[10]

In the end, it was Ruddick himself who settled the issue. Being more of a diplomat—and a much better person—than Georgia's racist governor ever was, Ruddick agreed to abide by the state's segregationist laws. "I don't care where I stay. Just give me a crowd of people and I'll have a good time," he said. "This might be a chance to open some people's eyes. I'm sort of taking this on an experimental basis. And who knows? Maybe it will help to make a better world for some people."[11]

And so it was that on November 18, a contingent of thirteen miners, their wives, and eighteen children—forty-four people in total—flew south from Halifax, bound first for New York and then on to the sunny south. Five of the rescued miners weren't well enough to travel yet, and one—Larry Leadbetter, the youngest of the crew—started out but returned home. He was still dealing with the lingering after-effects of a concussion he'd suffered in the Bump.

The Springhillers' plane had to land in Boston, and they spent a night there because New York—the connecting hub for the flight south—was shrouded in fog, and so it wasn't until the afternoon of November 19 that the excited travellers landed in Jacksonville, Florida. There they piled into two buses, and with a Georgia State

Patrol police car leading the way, they made the one-hour drive north to Jekyll Island. Greeting the Springhill visitors upon their arrival that Wednesday afternoon was a flypast by a US navy aircraft and a blimp with the words "Welcome guests from Nova Scotia" written on its side. A banner with the same message fluttered in the breeze above the entrance of the causeway that linked the island to the mainland. The message more truthfully would have read: "Welcome *white* guests from Nova Scotia." The reality was that Governor Griffin would rather that Maurice Ruddick, his wife, Norma, and the four children who accompanied them on the trip—the youngest ones in the family—had stayed home.

Twelve miners, their wives, and their children had rooms in a new but nondescript beachfront motel in the whites-only area of the Jekyll Island community. The Ruddicks were housed in a trailer three miles down the sprawling white-sand beach. To keep the Ruddicks company, Griffin's aides arranged for a Black academic, Dr. William K. Payne, and his wife, Mattie, to spend the week with the Canadians. Payne was the fifty-five-year-old president of historic Savannah State College, Georgia's first publicly funded Black post-secondary school. It's unclear what, if anything, a coal miner from Nova Scotia and an Alabama-born literary scholar were supposed to have in common other than skin colour. However, the Ruddicks and the Paynes got along well, and the cook, maid, and chauffeur who were provided to look after them saw to it that they wanted for nothing.

The white miners and their family members frolicked on the sandy beach, swam in the Atlantic, and were treated to buffet dinners, a dance, a boat cruise, and sightseeing and shopping bus trips. The Ruddicks were excluded from those activities, all of which were sponsored by local businesses. Instead, the Ruddicks and the Paynes went fishing, and they were guests of honour at barbecues, house parties, and a dance hosted by members of Jekyll Island's Black community. The locals even took up a collection that raised a hundred dollars so the Ruddicks would have some extra money to buy souvenirs for their kids.

Norma and Maurice Ruddick took their four youngest along when they visited Georgia, although three-year-old Jessie might have been happier to have stayed home. (Courtesy of Valerie [Ruddick] MacDonald)

Much to the dismay of Governor Griffin, a crowd of journalists, including a writer and two photographers from *Life* magazine, trailed the Ruddicks everywhere they went. Griffin, who initially had been delighted to have so much media coverage of the miners' visit, was chagrined both by the focus of the articles and their tone. Maurice Ruddick, "the Singing Miner," had stolen the spotlight. A writer for *Life* magazine described the Canadians' visit to Jekyll Island as "an impulsive gesture that wryly backfired on Georgia's Governor Marvin Griffin."[12] Sam Caldwell, the Griffin aide who'd instigated the Springhill miners' visit and invited the media to witness it, echoed his boss's lament. He complained to his wife, "The damn press outnumbers the miners. All they want is to file stories about this poor little old lonely Negro family at the far end of the beach deprived of all modern conveniences."[13]

Griffin was acutely aware of the tone of the press coverage and did his utmost not to be seen meeting with Maurice Ruddick. As *Globe and Mail* reporter Bruce West noted, no Georgia politician had ever been photographed with a Black man, and Griffin had no intentions of being the first.[14] His biggest fear was that a photographer would catch him unawares and snap a shot of him greeting Ruddick or even shaking hands with him. Happily for Griffin, he succeeded in avoiding either situation. While he briefly met with the Ruddick family and even deigned to shake Maurice's hand, the governor's aides made sure no photographers were on hand to record the moment. Later, when a reporter asked Griffin if he planned to meet with the Ruddicks, the governor smiled like a polecat as he explained he'd already spent time with the family. He said they were "nice people." Nobody believed he really meant that, but Griffin didn't care.

For his part, Maurice Ruddick never publicly volunteered his opinion of Georgia's governor. Whatever it was, he shared it only with his wife. When an interviewer asked Ruddick how he was enjoying himself in Georgia, he was diplomatic. "I seem to be enjoying myself just as much as the others" was all he chose to say.[15]

It was left to Bruce West to say what many Canadians were thinking when he commented that "Maurice Ruddick's good-humoured and robust reaction to the segregationist angle in the Georgia invitation is probably striking more blows against this distasteful situation than any number of solemn pronouncements. For one thing, he's making the rule on segregation look a little bit ridiculous. And if there's one thing that will make a bigot squirm every time, it's ridicule."[16]

THE SPRINGHILL MINERS and their families returned home on November 25, tired but happy. Their holiday in Georgia, brief though it was, had been one that none of the forty-four people who experienced it would ever forget. The trip had gone surprisingly well, all things considered. Most of the people they'd met in Georgia had been welcoming—all but Governor Griffin, who'd been pleasant in a way that a smarmy politician finds easy. The November weather had been sunny and warm. The accommodations had been comfortable. The sandy beach with its warm water was a paradise many of the Springhillers had previously glimpsed only in the movies. And what's more, the miners stepped off the plane in Halifax sporting new shoes given to them by the owner of a Georgia shoe factory. Then there were the job offers the men had to consider.

All the men who'd visited Georgia had been offered work at a seafood processing plant at St. Simons Island, 20 miles north of Jekyll Island as pelicans fly. Ultimately, only one man—Levi Milley—had any interest in moving south. Most of the others preferred to "play the waiting game." Hundreds of men were collecting a meagre weekly payout from the miners' relief fund while they waited. What they were waiting for, they had no idea. But Springhill was still "home," and they were loath to leave it.

More than two hundred out-of-work miners did opt to seek work elsewhere in Canada; Harold Brine was among them. While on his visit to Toronto, he'd received a job offer, and because his wife was eager to relocate there, the family did so. Unlike many of the other men who left Springhill, Harold Brine would never return home again.

Moving away wasn't as easy for Maurice Ruddick. With a dozen kids to feed and provide for, it wouldn't have been feasible for him and wife Norma to pack up and go. While they knew it would be difficult to stay put, they did so. They had their religious faith, their dreams, and each other to sustain them. The Ruddicks resolved to do their utmost to tough it out. Although doing so didn't earn him any money, Ruddick was kept busy at least for a few months answering his mail. He'd received more than four hundred letters from well-wishers near and far; the arrival in his mailbox of more envelopes each day raised his spirits and gave him hope. And hope in Springhill was an increasingly scarce—and precious—commodity.

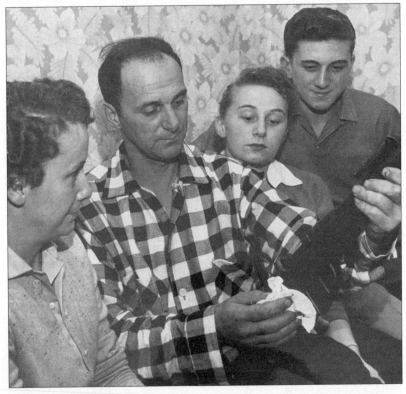

Times were lean for many Springhill miners following closure of the mine. Like many men, Gorley Kempt used his hunting rifle to put food on the family's table. In this photo, he checks his Enfield rifle as wife, Margie, daughter Betty, and son Billy look on. (Courtesy of Billy Kempt)

CHAPTER 28

LOOKING FOR ANSWERS

JANUARY 1959

LIKE MANY OF THE SPRINGHILL MINERS WHO NOW FOUND themselves out of work, town mayor Ralph Gilroy clung to the hope that better days were ahead. He had no alternative. At age forty-nine, he was too young to throw in the towel. Gilroy continued to believe that somehow, some way, his town would survive. He worked tirelessly to promote Springhill's economic potential, even though it was beginning to look increasingly likely that he was destined to be mayor of a ghost town.[1]

Some of the proposals that Mayor Gilroy and local business leaders considered for the revitalization of Springhill seemed to have merit, at least on face value. That was so with an offer made by an enthusiastic group of retired Toronto businessmen who volunteered to provide Springhill with pro bono advice and recommendations on how to attract new industries. Unfortunately, their efforts came to naught, as did many other similar initiatives.

Other ideas about how to save Springhill were unworkable from the get-go. Among them was the idea of using money from the miners' relief fund as seed capital for new local businesses, but this wasn't permitted under the terms of the fund's establishment. Then there was a pitch the miners' union made to DOSCO to reopen the shuttered No. 4 mine, albeit on a reduced scale of operations. With the entire Nova Scotia coal industry ailing, the bottom-line economics of reopening any of the Springhill mines were untenable, and so although that idea was soon shelved, it didn't dissuade a group of unemployed miners from trying to start a bootleg mine. There was no shortage of men willing to go back to work underground. Miner Herb Pepperdine, one of the Seven, was eager to be among them if that ever happened. As he told an interviewer, coal mining work was all he knew. If it hadn't been for the tragic events of October 23, 1958, in all likelihood he'd have worked in the mine another twenty-four years, until his scheduled retirement date in 1982 at age sixty.[2]

The most bizarre suggestion put forward for new businesses in Springhill called for setting fire to the abandoned underground coal seams. The hope was that doing so would produce a perpetual supply of gases and steam that could be harnessed for low-cost electrical generation. Never mind the danger of uncontrolled carbon emissions in a subterranean blaze that would smoulder perpetually deep in the earth beneath Springhill. In 1958, any worries about carbon emissions and climate change were still a half-century in the future for most people.

"The only thing left now is for the federal and provincial governments to take over," said a frustrated Mayor Gilroy. "I don't know what they can do, but I'm hopeful they can keep Springhill alive."[3] That would prove to be more easily said than done. Once the townspeople had buried their dead and dried their tears, they began thinking about the future in a one-industry town in which that industry was now closed. Predictably, the mood on the streets of Springhill darkened.

For the first few weeks after the disaster, the pages of the *Springhill-Parrsboro Record* were full of articles praising the tireless efforts of various local citizens' committees, civic leaders, and the provincial government. All of them continued their valiant struggle to convince new industries to set up shop in town. For his part, Ralph Gilroy was busy travelling back and forth to Montreal, Ottawa, and Toronto promoting awareness of the disaster relief fund and spreading the "There Will Always Be a Springhill" message. Sadly, the mayor was fighting a losing battle.

While donations to the miners' relief fund continued to come in and eventually exceeded $2 million, the efforts of all those advocates who were touting Springhill as an ideal community in which to do business mostly came up empty. As the Canadian Press reported on the first anniversary of the Bump, the drive for new industry "has produced only a small woodworking plant employing about fifteen. A prison farm will give jobs to some older men. A proposed battery factory may employ fifty."[4]

And so it was that a slow, steady exodus from Springhill of miners and their families continued. The town was withering, its population shrinking. More and more abandoned houses were sitting empty. Businesses found it harder and harder to stay open. School enrolments dropped, and Springhill's municipal tax base was eroding. Mining had always been the town's lifeblood, and now with the last mine closed, it was difficult to see a way forward. In that regard, Springhill wasn't alone in Nova Scotia. The province's coal mining industry was ailing. It had been for several years.

Speaking in Toronto at an October 1953 meeting of the Empire Club, the pro-British business lobby group, then-DOSCO president Lionel Forsyth had sounded the alarm. "Our coal operations in Nova Scotia have been for some years beset by steadily increasing costs of production, lower productivity of man power and intensified competition from hydro-electric power, oil, and natural gas,"[5] he said.

By the late 1950s, it was common knowledge that all DOSCO's coal mines were facing gale-force competitive headwinds. They'd come to rely too much on subventions—government subsidies that reduced rail and water shipping costs—at a time when demand for coal was shrinking and production costs were rising. Even more problematic was that American mines were now producing ever-greater volumes of coal at a cheaper cost than Nova Scotian mines could. DOSCO, unable to sell more than a million tonnes of coal, announced in January 1959 it was closing its Cape Breton mines for the entire month of February, a move that idled four thousand miners. After that, the company revealed it planned to reopen the mines only on a part-time basis until the end of May. "We've reached the very limit. The markets [for Nova Scotia coal] just don't exist," said DOSCO vice-president Harold Gordon.[6]

The problems that Nova Scotia's coal industry faced were so significant that Premier Robert Stanfield led a delegation of provincial and municipal government officials, DOSCO managers, and miners' union leaders to Ottawa for a meeting with Prime Minister John Diefenbaker. Afterward, although nothing of any lasting significance came out of Stanfield's efforts, the premier did his best to play down the significance of the disappointment. He insisted he wasn't discouraged. However, many other people clearly were, and they didn't hide it.

There was a lingering sense of despair in Nova Scotia mining communities and a growing demand for answers to some of the life-and-death questions that the province's coal industry faced. Despite the fact the people of Springhill had already lost their mines, they were eager for a public inquiry into the Bump that had devastated the town. They hoped a probe would shed light on the cause of the events of October 23 and that it might determine whether the contentious mining practices DOSCO had pressed ahead with at the mine the previous spring—when the three coal faces had hurriedly been lined up—had played a role in triggering the upheaval or increasing its intensity. Seventy-five men were dead, and the town's future had been thrown into doubt.

Springhillers wanted to know why, and human nature being what it is, they wanted someone to be held accountable.

ROYAL COMMISSIONS AND public inquiries are always easy targets for criticism.

In Canada, such proceedings investigate matters of national concern, usually an important general issue or a specific incident that has occurred. Appointed by the Governor-in-Council—that is, the Cabinet—they include a panel of distinguished individuals, experts, or judges. An order-in-council sets out the terms of reference for the commission as well as the powers and the names of the commissioners. Federal inquiries are generally limited to matters within the constitutional jurisdiction of the Parliament of Canada, while provincially appointed commissions and public inquiries look only at matters over which the provinces have control.

In the 156-year history of Canada, there have been about 450 federal commissions of inquiry with and without the royal title. No one has counted the number of such provincial inquiries. No matter. It's fair to say that governments in this country make liberal (with a small *l*) use of royal commissions and public commissions. They rival hockey, talk about the weather, and complaining as Canadians' national pastime.

Skeptics opine that these costly proceedings are staged mainly to give an embattled government a way to do nothing while appearing to be taking decisive steps to resolve a controversy or get to the bottom of a befuddling problem. Whenever hearings are lengthy and complex or proceedings drag on—for weeks, months, and even years—the public tends to lose interest, tempers cool, and any crisis that's been building stands a good chance of being averted. None of that happened when the Stanfield government announced a public inquiry into Springhill's latest mine disaster.

On January 26, 1959, a three-man commission headed by prominent Halifax lawyer Donald McInnes began five days of well-attended hearings at the town's Legion Hall. Also sitting on the commission panel were Calgary mining engineer Harry Wilton-Clark and Thomas McLachlan, a former miners' union official from Glace Bay. Their mandate was to delve into all aspects of the October 23 disaster in hopes of discovering its probable origin and causes—and indeed those of all the bumps that had rattled the mine. The Stanfield government had also tasked the commissioners with determining whether the mining practices DOSCO had been using in the No. 2 mine had caused or contributed to the disaster, and whether the mine had been operated in compliance with applicable federal and provincial regulations.

Two of the commission members toured those parts of the No. 2 mine that were still accessible, while all three commissioners heard testimony from thirty-seven witnesses. Among them were all the DOSCO managers who were centrally involved in the mine's operations and in the post-Bump rescue and recovery operations. Harold Gordon, Louis Frost, and George Calder were among them. Also giving testimony were miners' union officials; miners who'd been working on the evening of October 23; and federal Department of Mines employee Herbert Zorychta, the resident engineer at the No. 2 mine from 1954 to 1958 and the initiator of extensive studies of the stresses that had been exerted on the mine's coal faces.

The McInnes Commission's final report included two hundred pages of verbatim testimony, background information, and the commissioners' findings; recommendations on the mine's operations that were put forward by the various interested parties; and a wealth of appended material. The entire document makes for interesting reading.[7] However, there's nothing in its conclusions or final recommendations that's revelatory or unexpected. This had little to do with the work of the commissioners, their staff, or the evidence the witnesses provided. Rather, it had everything to do with the task the provincial officials had given McInnes and his associates.

The essence of the commission's mandate was to sift through a mountain of data and conflicting opinions and come up with some definitive answers. That was impossible to do. There was no smoking gun, no villain to be denounced—other than the No. 2 mine or the coal mining industry itself. McInnes acknowledged as much when he wrote, "The Commission readily recognizes that it has been unable to reach any conclusive finding as to the cause or origin of bumps. A great many factors and circumstances have to be considered. The tremendous pressures which arise when mines are deep in the earth bring about circumstances concerning which there is little available information."[8]

That wasn't entirely true. More than two decades later, in 1980, a graduate student at Queen's University in Kingston, Ontario, would write a Ph.D. thesis about the 1958 mine disaster at Springhill. English-born and -educated, Keith Notley was a mature student who'd worked in Sudbury nickel mines for more than a decade and had conducted cutting-edge research into the then new science of rock mechanics before immersing himself in academia. Notley was casting around for a suitable topic for a doctoral study when Peter Calder, head of the Mining Engineering Department at Queen's, suggested he explore the causes and effects of the Springhill mine disaster. Calder had a compelling reason for doing so. The Springhill native was the eldest son of DOSCO mine manager George Calder, and he had long wondered what caused the Bump. Calder thought Notley might be the ideal person to investigate. As it turned out, he was.

While Notley ultimately was unable to pinpoint the exact cause of the October 23 bump—which he concluded was "a natural consequence of mining at great depth in comparatively weak rock"—some of his other findings were revealing. For one thing, his inquiries led him to conclude that the miners were justified in worrying about the decision by Louis Frost and the coal company's other senior managers to line up the three longwall faces where coal was being extracted. The data suggested this well might have played a role in triggering the Bump and in fuelling its intensity. What's more, Notley's painstaking analysis of the stress and load data collected by the provincial Department of Mines officials who'd been studying bumping problems at Springhill for years prior to the Bump revealed some intriguing findings.

Making use of computer technology that wasn't available in 1958, and drawing upon his own technical expertise, Notley was able to decipher and manipulate the wealth of displacement data that had been recorded by monitoring equipment in the mine. When he did so, he discovered that in the six weeks prior to the Big One, not only had the incidence of bumps that rattled the No. 2 mine been growing, but so too had the bumps' intensity. A chart Notley created showed that warning signs were peaking on Thursdays and Fridays each week. By late October, the number of intense bumps had jumped dramatically, and the upward trend in this data was ominous.[9] Hindsight being twenty-twenty, it now seems evident that if only Louis Frost and his colleagues had been able to decipher the raw data in the same way Keith Notley did two decades after the fact, they might well have had reason to pause work in the No. 2 mine. Or even to reconsider its long-term future. However, that's all "what if?" conjecture.

In October 1958, although people had their suspicions, no one really knew how serious the threat of a major bump really was. Many of the veteran miners and the leaders of their union were fearful the danger was real and growing; their gut instincts told them so. Louis Frost had a different view. He was convinced the miners' fears about the impact of lining up the coal faces were baseless, and the company wasn't prepared to revert to having staggered coal faces. For a variety of reasons, it simply couldn't afford to do that. With the mine's economic viability already in doubt, so too was the future of Springhill itself. Everyone knew that closing the mine would kill the local economy and the town itself. If ever the expression "damned if they did and damned if they didn't" applied to a situation, it was the one in Springhill, Nova Scotia, in the autumn of 1958.

INFORMATION ABOUT HOW and why bumps happened wasn't the only aspect of the Springhill mine disaster that attracted the attention of scientific researchers. In 1958, the Cold War was at its peak, and fears of a nuclear clash between the United States and the Union of Soviet Socialist Republics (USSR) were running high. This was the period when the Canadian and American governments were building fallout shelters for government and military leaders and were encouraging citizens to build their own shelters in the basements of their homes. However, there were myriad questions and many unknowns about what might happen if people really had to live for prolonged periods—weeks, months, or even years—cooped up in a sealed underground bunker. Surviving a nuclear blast was one thing; staying sane afterward was quite another.

South of the border, various government agencies of the US federal government were studying this issue. When news reports about the experiences of the Springhill miners who'd been trapped underground were broadcast in the United States, officials of the National Academy of Sciences–National Research Council (NAS–NRC) in Washington saw an opportunity for an interesting case study. Both Carl Pfaffmann, the chairman elect of the Anthropology and Psychology Division of the NAS–NRC, and Neal Miller, its past chairman, were psychiatrists, and both were interested in learning more about the effects of isolation on those who experienced it. They also were keen to study how and why leaders emerge in such situations. Remember, the Springhill mine disaster happened just four years after publication of British author William Golding's now classic novel *Lord of the Flies*[10] and one year after Nevil Shute's post-apocalyptic novel *On the Beach*.[11]

In the shadow of a potential nuclear holocaust, there was no end of interest in the late 1950s—both among the general public and in the scientific communities—in the topic of how post-disaster isolation might affect group dynamics of people who survived any such disaster. With that in mind, officials in the NAS–NRC's Disaster Research Group recruited a team of five Nova Scotia academics—two psychiatrists, two psychologists, and a sociologist—to conduct a wide-ranging study of the experiences of the trapped Springhill miners in hopes of learning how those experiences had affected them. "The emphasis [was] on the men who were trapped—their behavior during entrapment, their physical and particularly their psychological state after rescue."[12]

The researchers interviewed seventy-six people, including all nineteen of the trapped miners and seventeen of their wives. Springhill was dubbed "Minetown" in the study, and the names of all the individuals who took part were withheld to protect their identities and ensure they'd be candid in their replies to the interviewers' questions. However, if you read Disaster Study No. 13 (which the lead authors, Professor H.D. Beach, a Dalhousie University psychologist, and Professor R.A. Lucas, an Acadia University sociologist, published with the NAS–NRC in 1960), it's easy to figure out the identities of some of the men. For example, "D6" who is described as "a member of the minority Negro community in Minetown" is readily identifiable as Maurice Ruddick. And anyone who's familiar with what happened in Springhill can easily identify several of the other interview subjects. American author Melissa Fay Greene, who had the opportunity to personally interview several of the trapped miners as well as some of their wives and relatives, cracked the academics' code. In doing so, she determined who was who in the study and used that information to write a 2003 book about the disaster that chronicled the horrific details of the underground ordeal endured by the two groups of miners who were rescued.

As for the study itself, the researchers found some intriguing parallels in how both groups of trapped miners—the Seven and the Twelve—passed the time. In the first three days, "the escape period," individuals who were energetic, resourceful, and handy with tools led the way in efforts to find a way out of the mine; they included Gorley Kemp, Garnet Clarke, and Harold Brine. However, during the "survival period" that followed, older men who were more measured in their actions and who were group-oriented gradually assumed leadership roles and kept things together; the two who stood out in doing so were Caleb Rushton and Maurice Ruddick, both of whom were mature men of religious faith.

Disaster Study No. 13 is fascinating reading; however, the authors' bottom-line findings were ultimately inconclusive since they "were derived within a situation whose essential features may be relatively unique and they are based on data from rather small incidental samples." That said, the authors hastened to add that the information they'd presented was "amenable to further investigation within social science research as well as within the special field of disaster research."[13]

IT WASN'T ALL academic studies and bad news for the residents of Springhill following the closure of the No. 2 mine. Millions of people across North America and around the world had heard about the courage of the town's residents and followed media reports of the heroic rescue operations that resulted in "the miracle at Springhill." Others learned of the Springhill mine disaster when American folk singer Peggy Seeger wrote a song she called

"The Ballad of Springhill" (which the people of the town would come to adopt as an unofficial theme song). Seeger had never been to Springhill; she'd been inspired to take pen in hand after viewing the CBC television coverage of the disaster as she sat in a Paris café. Her tune was popularized and reached a mass international audience when it was sung by the well-known 1960s folk revival group Peter, Paul and Mary.

Many people in Canada and the United States had also learned about what happened in Springhill after seeing *The Ed Sullivan Show* that featured the appearance by Gorley Kempt, Caleb Rushton, and Dr. Burden, or else they'd read newspaper editorials that lauded the courage of Springhillers. Thousands of individuals contributed money to the miners' relief fund. Then too, there were the honours and awards that civic-minded organizations and the media bestowed upon the townspeople.

In November, just days after "the miracle at Springhill," the Royal Canadian Humane Association (RCHA) awarded its Gold Medal for bravery in lifesaving to the people of Springhill—especially the barefaced miners and draegermen who'd risked their lives to rescue trapped miners and recover bodies from the devastated No. 2 mine.[14]

Then, just a few months after the announcement of the RCHA award, an even more prestigious honour came the way of the people of Springhill. This one was internationally renowned. The Pittsburgh-based Carnegie Hero Fund Commission announced on May 26, 1959, that the 379 miners, mine officials, doctors, and other rescuers who'd been involved in the Springhill mine disaster were being collectively honoured with a Carnegie Medal in recognition of their heroism. The gold medal, which was inaugurated in 1904 by Scottish-American industrialist-philanthropist Andrew Carnegie, honours deeds of heroism in which men or women are injured or die while trying to rescue others. According to a statement issued by the Carnegie Commission, this was only the second time in the award's fifty-five-year history that a group had received it. The first was in 1912 when the passengers and crew of the *Titanic* were so recognized. "The courage and determination of [the Springhill] men who were willing to sacrifice their lives that others might live exemplified the highest degree of heroism to the people throughout the world," the Carnegie citation stated.[15]

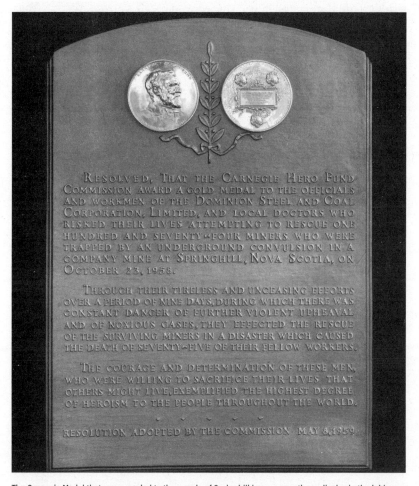

The Carnegie Medal that was awarded to the people of Springhill is permanently on display in the lobby of the Dr. Carson & Marion Murray Community Centre on Main Street in Springhill. (Carnegie Hero Fund Commission)

News of the Carnegie Medal was most welcome in Springhill, but it was yet another award—this one individual—that set tongues wagging. It also hinted at an undercurrent of racism that had always existed in Springhill; like an iceberg, the ill will was mostly below the surface and out of sight. However, it revealed itself in mid-January 1959 when the *Toronto Telegram* polled its readers on their choice for the newspaper's Citizen of the Year Award. There were twenty-one nominees—Conservative prime minister John Diefenbaker reportedly being among them.

Heroism, like beauty, is in the eyes of the beholder. *Telegram* readers had read the newspaper's extensive coverage of the Springhill mine disaster and had concluded that Maurice Ruddick, "the Singing Miner" from Springhill, Nova Scotia, was a hero who deserved to be lauded. He had garnered fifty-one percent of the thousands of votes that were cast. As the January 22 announcement of the winner explained, "The 46-year-old father of 12 children was named by the voting thousands not only for his underground bravery; they cited, too, his generous tolerance when he, a Negro, accepted segregation on the subsequent Georgia vacation, so that his mates would not miss the trip."[16]

Ruddick himself was as surprised as he was delighted by this latest honour. He learned about it in a phone call, which he had to take at a neighbour's house. The Ruddicks didn't have a phone of their own; they couldn't afford one. Maurice and Norma travelled to the *Telegram* award ceremony in Toronto. While they were there, Ruddick's photograph appeared on the front page of the newspaper yet again. He also was lauded by Ontario premier Leslie Frost and appeared as a mystery guest on the popular CBC television program *Front Page Challenge*.

If Ruddick hoped that people back home in Springhill would be pleased that one of their own was being honoured in Toronto and that he was bringing attention to the town, he was sadly mistaken. It made no difference to them that Ruddick had made a point of saying he was accepting the *Telegram* award on behalf of "all the community of Springhill." More than a few of the Ruddicks' neighbours—and even some of the miners with whom he'd been trapped—complained that, as they saw it, the media had singled Ruddick out for praise and built him up as a hero because he was Black and because he was "the Singing Miner." Others carped that he'd gone out of his way to trumpet his own praises.

The truth, as is so often the case, undoubtedly was to be found in the quiet eye that is at the centre of most storms of controversy. It was Gorley Kempt's wife, Margie, who gave voice to what many Springhillers doubtless knew to be true when she said, "We were all used as a marketing gimmick. A gimmick in Georgia, and a marketing gimmick back in Canada."[17]

IN THE WAKE of the No. 2 mine's closure, many miners and officials from the miners' union pushed for the reopening of at least one of Springhill's shuttered DOSCO mines. There were also efforts to further increase production at the bootleg mine that had started up, but all such operations were short-lived and doomed to failure for a variety of reasons, both economic and logistical.

Work crews spent several months combing through the ruins of the No. 2 mine, salvaging any machinery and tools that could be repurposed in the DOSCO mines still operating in Pictou County and Cape Breton. After that, the No. 2 mine sat abandoned and derelict for about a year before it was sealed shut. Peter Calder recalled, "People from *Reader's Digest* magazine contacted my father, or maybe it was someone at the Dominion Coal Company. The *Reader's Digest* people wanted to send a writer into the mine. On the day in July 1959 that it was to be closed, I went down into the mine with my father and with the *Reader's Digest* writer. We toured the levels and saw the spots where miners had been trapped and had managed to survive.

"My dad and I were the last two people ever to be inside the No. 2 mine. When we came up after spending several hours down there, it was like a military operation. There was a large group of workers involved, and five or six cement trucks were standing by at the top of the incline shaft. As soon as we emerged, they started pouring a huge cement collar around the mine's entrance. Nobody has ever entered that mine again."[18]

Epilogue

For eighty-five years, the coal mines that ran deep under Springhill were both the lifeblood and the curse of this hard-luck town. Today, those same tunnels are flooded with geother-mally heated groundwater that's drawn from shallow parts of the old mine into heat pumps on the surface (where it arrives at a constant temperature of about 64 degrees Fahrenheit) and then returned to the subterranean depths. The clean, low-cost renew-able energy that's generated in this process is being used to heat and cool businesses and commercial buildings in Springhill's bur-geoning geothermal business park.[1]

Two hundred yards to the west, on the former site of the sprawl-ing mine complex that was once the beating heart of the town of Springhill, you'll now find only some scrubby trees, a bumper crop of weeds, emptiness, and the ghosts of yesteryear—among them the 424 men who died in the Springhill pits. All the mine buildings that once stood here are gone. They've been demolished. The pit-heads have been sealed shut; the railway tracks have been torn up; and even the wash house and light cabin have been demolished. The only physical reminders of the past are two old mine trolley cars that sit on the verge of the roadway they now call Miners

Memorial Drive, and a couple of cairns on which are mounted National Historic Site commemorative plaques.

At the time of the Bump, in October 1958, the town of Springhill was a proud, vibrant community of more than seven thousand people. In the wake of the mine disaster, many families left town; few ever returned. "Goin' down the road" was the familiar expression used by Maritimers at the time when they moved to central Canada (usually to Toronto) or farther west looking for a fresh start. In one of the strangest twists on this theme, the well-meaning mayors of several Canadian cities in central and western Canada—Don Mackay from Calgary being prominent among them—offered to "adopt" Springhill families, especially those consisting of a widow and children whose breadwinner had died in the Bump. How a single mother with children in tow would ever build a new life in a new city was never made clear, but never mind that; many well-intentioned offers to help with relocating were made. One woman who received such an invitation was Tommie Tabor's thirty-eight-year-old widow, Ruth.

"Mom suddenly was on her own with four children to raise," said Valarie (Tabor) Alderson. "My mother appreciated the Calgary mayor's kind offer, but she wanted to stay in Springhill, where she had family. But one of her brothers, Tom McManaman, got the same offer. He went west and stayed, lived there for the rest of his life."[2]

Among the other Springhillers who moved away and never came back was Harold Brine, one of the four men who have been spotlighted in these pages. After relocating to Toronto in 1958, he stayed there for four decades while working at a variety of jobs. His first marriage ended in 1973; he and his second wife, Murriel (née Campbell), married in 1982, and in 1999 they moved back east to settle in Geary, New Brunswick, where they still live. When Harold celebrated his ninetieth birthday on June 25, 2022, he was the last survivor among the lucky men who cheated death in the shattered mine and were rescued by fearless barefaced miners.

In this 2018 photo, Harold Brine, who was eighty-six at the time, holds a picture of himself taken when he was in hospital after his rescue from the mine. He carried the photo in his wallet for more than sixty years. (Courtesy of Wally Hayes)

Brine's co-worker Maurice Ruddick lived out his life in Springhill. "I can't leave this town," he told a magazine journalist who interviewed him.[3] Unlike so many of their neighbours and friends, Maurice and Norma Ruddick stayed put, although doing so was anything but easy, especially after they'd made it a baker's dozen of children, adding a thirteenth child to their family in 1960. It was now that Norma, who by then was thirty-five, announced that was it: no more babies.

Maurice remained ever hopeful that some new employment opportunities would arise in Springhill, but doing so was akin to waiting for Godot. Those new jobs never came. To heat the family home in winter, Ruddick picked loose coal wherever he could find it along the old railway siding and on the abandoned DOSCO mine property.

Jobs were scarce for a middle-aged man of colour in Cumberland County. Until 1962, Ruddick made do on thirty-five dollars per week he received from the Springhill Disaster Relief Fund (which inevitably ran dry). He then scratched out a living by collecting thirty-dollar-per-week welfare payments, toiling at odd jobs, and earning five dollars per performance whenever he sang with his five youngest children, "the Singing Miner and His Harmony Babes." When he applied for a job as a guard at the new federal prison farm that opened in 1967—a facility he'd helped build—Ruddick was turned down, supposedly because of his age. He was fifty-four at the time and was destined to live another twenty-one years.

Unlike many of the miners he'd worked alongside, Maurice Ruddick continued to live in Springhill and to hope in vain that the mine would reopen and he could go back to work there once more. (Clara Thomas Archives and Special Collections, York University)

Maurice Ruddick died on June 25, 1988. The great irony about his final years is that while he was a forlorn figure, of the nine hundred men who were working in the No. 2 Springhill mine in 1958, and the nineteen men who were rescued in "the Springhill miracle," he's the only one who's remembered today by people outside Cumberland County. "The Singing Miner" is the subject of a Historica Canada Heritage Minutes video[4] and of a critically acclaimed 2011 stage play. *Beneath Springhill: The Maurice Ruddick Story*, written by Peterborough, Ontario, playwright-actor-singer Beau Dixon, has been performed at theatres across Canada. "Maurice Ruddick was a man who cared deeply about his wife, his family, and his community," said Dixon. "He went to his job in the mine every day because he had to put food on the table and thought of others before he thought of himself. To me, he was a perfect example of a blue-collar working man, and he was a hero."[5]

A third Springhiller who figures prominently in the story of the 1958 mine disaster is Dr. Arnold Burden. Unlike so many other people, he and his wife, Helen, remained in town after the Bump. The good doctor continued to practise medicine until he retired in 1993 at age seventy-one. When he passed in March 2018 at the age of ninety-five, an obituary that appeared in the Canadian Medical Association newsletter noted that "the devastating impact on the community [of the 1956 and 1958 mine disasters] led to him staying in Springhill to help his community rebuild."

Such efforts proved to be painful for Burden and for all who stayed in town. By 2014, with its tax base continuing to shrink, it had become increasingly problematic—if not impossible—for the remaining 2,700 residents of Springhill, a quarter of whom were seniors, to provide and pay for basic public services. As a result, after 126 years of incorporation, the mayor and town council made the painful decision to dissolve the town and became part of the Municipality of the County of Cumberland. But even so, the pride in community and the spirit that sustained the people of Springhill live on. They will never die.

SPRINGHILL ISN'T THE only coal mining town in Nova Scotia to have suffered through the loss of its coal mine and the painful transition period that followed. The last colliery in Sydney Mines closed in 1975, and the last mine in Glace Bay ceased operation in 1984. Then there was the ill-fated Westray mine that opened in Plymouth, Nova Scotia, in September 1991 with funding help from the provincial and federal governments. But despite high hopes for the initiative, Ol' King Coal's restoration to his Nova Scotia throne was short-lived. The Westray mine closed after just eight months after the horrific May 9, 1992, underground methane explosion killed twenty-six miners and sealed the mine's fate. There would be no Westray miracle and no happy ending for the miners, their loved ones, or the 117 mine employees who suddenly found themselves unemployed; what was even worse, they were forced to battle to get severance pay and the other money they were owed. It was a sad, disturbing story, much like that of Nova Scotia coal mining in general.

At various times over the last two centuries, there were collieries in seven of Nova Scotia's eighteen counties. More than three hundred underground mines have operated since commercial mining began in Cape Breton in 1720, and guesstimates are that as much as 500 million tonnes of coal were mined in Nova Scotia's surface and underground coal mines. Now all those mines are history—save for the controversial colliery that reopened at Donkin in September 2022.[6] Officials in the province's Department of Natural Resources and Renewables have created an Abandoned Mine Openings Database.[7] It includes references to more than seven hundred abandoned mine workings—with 8,400 shafts, adits, slopes, open cuts, trenches, and associated underground workings—at both underground operations and advanced exploration sites. And those environmental scars and hazards aren't the only bitter legacy of two centuries of coal mining.

Arnold Burden celebrated his ninetieth birthday on April 29, 2012. (Courtesy of the Burden family)

Over the years, at least twenty-five hundred men have lost their lives in Nova Scotia's coal mines.[8] Beyond that are the thousands of others who fell ill with silicosis, emphysema, and a host of cancers and heart ailments related to a lifetime of working underground.

You might be surprised to learn that while commercial coal production in Nova Scotia is all but finished, as late as 2023 Nova Scotia Power continued burning coal to generate forty-eight percent of its electricity. "[It] accounted for upward of half of Nova Scotia's greenhouse gas emissions in 2005," the *Globe and Mail* has reported. "Since then, emissions from electricity are down by more than a third, with emissions economywide also down by more than a third."[9]

There's more good news in this regard. Nova Scotia has plans to tap into the Maritime Link—the new $1.7 billion transmission connection with Newfoundland and Labrador's new Muskrat Falls hydro dam. If all goes as hoped, by 2026 this project will supply as much as thirty percent of Nova Scotia's power needs, while another thirty percent will come from wind and solar.

Some of the coal Nova Scotia Power has been burning—at least until 2020—was mined in the Stellarton and Cape Breton regions of the province. Almost all of what was being burned after that reportedly came from the United States and Colombia. Importing has been the primary source of coal for more than twenty years. As for the quantity of coal purchased, fuel purchasing information is considered commercially sensitive, and so Nova Scotia Power wouldn't share the number publicly.

As an aside, just as Springhillers whistled past the proverbial graveyard in hopes the disaster everyone knew was inevitable somehow wouldn't happen, much of the world continues to burn coal. A dismaying amount of it. We just can't help ourselves. Humanity remains deaf to the jangling alarm bells warning us of impending climate disaster. According to information posted on the website of the World Coal Association—whose members include "global corporate companies and not-for-profits with a stake in the future of coal"—in 2023 thirty-seven percent of the world's electricity and more than seventy percent of the world's steel were still produced using coal.[10]

Canadians have no reason to feel smug. We continue to be part of the problem. This country is still a coal producer—in effect, a dealer for the junkies. According to the latest figures posted on the coal industry website NS Energy, Canada ranks thirteenth globally in coal production, with an estimated annual output of 57 million tonnes. Not surprisingly, China is the world leader both in coal production and usage, burning 3.7 billion tonnes annually; that's forty-seven percent of the world's total, and it accounts for a huge percentage of greenhouse gas emissions. India is a distant second at 783 million tonnes, while the United States is third at 640 million tonnes.[11]

British Columbia and Alberta are the two Canadian provinces that produce most of the coal mined in this country. Two-thirds of it is exported, mainly to Asia.

WHILE OL' KING COAL no longer reigns in Nova Scotia or in Springhill, Canadians continue to remember that star-crossed community for a couple of reasons. Of course, they know it as the birthplace of singer Anne Murray (and the home of the Anne Murray Centre, which has been a popular tourist attraction on Main Street since 1989). And Canadians also remember Springhill as the site of the epic 1958 mine disaster. Although there were three such calamities, it's the 1958 bump that remains top of mind for Canadians, especially those of a certain age. There are many reasons for that, not least of which is the television coverage of "the miracle at Springhill," when fearless barefaced miners plucked those nineteen men from would-be graves in the shattered mine.

In Springhill itself, the community's coal mining legacy endures in the memories and in the bloodstreams of a dwindling number of older residents. It's also there for the world to see in those National Historic Site plaques that mark the location of the old mine. It's in the town's splendid miners' museum.[12] It's there in the hundreds of headstones that dot the windswept Hillside Cemetery to the east of town, and it's there in the miners memorials that stand in the park at the top of the Main Street hill—beside the old Miners Hall, which now serves as the town library.

319

Three of the eldest Ruddick daughters—(*l-r*) Ellen, Sylvia, and Valerie—sang at an October 23, 2008, commemorative event held at the Dr. Carson & Marion Murray Community Centre in Springhill to mark the fiftieth anniversary of the mine disaster. (Courtesy of Wally Hayes)

Soon, you'll also be able to see a colourful reminder of the town's mining heritage in the Blood on the Coal commemorative tartan that seventy-two-year-old Springhill-born artist and civic activist Roberta (née Ward) Bell has created and registered with the Scottish Register of Tartans. Bell was only seven at the time of the Bump, but she hails from a coal mining family, and she's never forgotten how the disaster affected her and her hometown. "The idea for a tartan came to me in a dream one night in 2019," said Bell. "When I woke up, I wrote down exactly what I thought the tartan should look like."[13]

In her vision, the miners' tartan was hued tricolour—black for coal and the darkness of a mine, yellow for the light that comes from miners' lamps, and red for the blood of the thousands of men in Nova Scotia who lost their lives or were seriously injured working underground.

"I feel that the idea for a tartan was a gift that was given to me, and I had to do something with it," said Bell.

Her hope is that one day soon, everyone in Springhill and other coal mining towns will be wearing Blood on the Coal tartan garments—shawls and scarves for ladies, and ties, kilts, and tams (or whatever other garments are created) for men. The black, red, and yellow of the tartan will provide a fitting colour scheme for the annual Miners Memorial Day events held each June 11. That's when people in Springhill and in Nova Scotia's other former coal company towns gather to reflect on the province's coal mining heritage and to honour and remember all those miners who died on the job and as a result of workplace-related chronic health conditions and debilitating injuries.[14]

Said Roberta Bell, "Now that coal mining is a memory, it's important to remember those men and boys who went into the mines every day, as well as the women who endured the hardship of a miner's life to put food on the table and provide for their families."

Nowhere in Nova Scotia—or in Canada—is it more fitting to do so than in Springhill, Nova Scotia, a community in which there truly was blood on the coal. There's a lesson in the story of the Springhill mine disaster that's both poignant and profound. That is, if only we're wise enough and willing enough to pay heed.

ACKNOWLEDGEMENTS

THE BRITISH AUTHOR NEIL GAIMAN ONCE OFFERED A BIT of whimsical advice on how to write a book. "This is how you do it," he explained. "You sit down at the keyboard and you put one word after another until it's done. It's that easy, and that hard."[1]

Gaiman had it right. There's nothing mysterious about the writing process. However, because he's a fiction writer, he neglected to mention one important point: for writers of non-fiction, the creative process is a tad different. The words a novelist commits to paper are the product of his or her own imagination, while those of a writer of non-fiction are the fruits of conversations between the writer and those people who have provided information or have shared their personal memories of people and events.

In a time of pandemic, when travel was curtailed and arranging meetings with sources at times was difficult, the task of writing this book was challenging in ways I'd never envisioned. For that reason, I could never have completed it without the help, encouragement, and kindness of a great many people.

I owe special thanks to my editor at HarperCollins (Canada), Jim Gifford, copyeditor Patricia MacDonald, and their now retired colleague Patrick Crean; it was Patrick who nurtured my idea for a book on the Springhill mine disaster and who encouraged me to forge ahead with the project. A huge thank you also goes to my agent, Richard Curtis in New York. Richard has guided and counselled me with sagacity for more than two decades; I'm forever in his debt.

As you'll know if you've read this far, this book focuses on the experiences of a trio of individuals whose lives were profoundly affected by the Bump and on one man who died when the earth heaved. I want to applaud the following individuals for helping me put human faces on each of these men when they recounted details of the tragic events and the miracle rescues that were part and parcel of the remarkable story of the Springhill mine disaster.

Harold Brine, at age ninety-one in 2023, is the last survivor among the nineteen rescued miners. Harold was my "go-to guy" for all questions about himself (naturally), the Springhill mine, and the Bump. This Geary, New Brunswick, resident and his dear wife, Murriel, were infinitely patient as they answered what surely must have seemed to them like an endless stream of silly questions. However, they did so with grace and unfailing good humour. Not only were they invaluable sources, I'm honoured to say they have become good friends.

Valerie (Ruddick) MacDonald, the Ruddick family historian, has helped me beyond measure. Val and her husband, Barry, who live in Riverview, New Brunswick, took the time to travel to and then show me around Springhill, introduced me to many townspeople who recounted their memories of the Bump, and provided me with a wealth of information. Val also answered my questions about her dad, provided me with splendid archival photographs—some of which are included in the pages of this book—and opened my eyes to the at times harsh realities of what it means to be a Black person in Canada. Like Harold Brine, Val has become a good and valued friend.

Springhill resident Valarie (Tabor) Alderson, the daughter of

Raymond "Tommie" Tabor, shared her memories of her dad and of those dark, painful days that followed his untimely death and the closure of the DOSCO mine. I thank Valarie Alderson for her kindness and for bringing her father's memory to life for me; he was one of the brave miners who paid the ultimate price while toiling to put food on the table for themselves and their loved ones. There were seventy-four other men who suffered the same cruel fate.

Windsor Junction, Nova Scotia, resident Bill Burden, the eldest of the four sons of Dr. Arnold Burden and his wife, Marion, provided me with information about his dad's life and medical career and about life in Springhill at the time of the mine disaster. Two of Bill's younger brothers, Tim, who lives in Dartmouth, Nova Scotia, and David, who resides in Sarnia, Ontario, were also kind enough to offer their insights, feedback, and words of encouragement.

In addition to all those who provided me with detailed information on the individuals I've spotlighted in this book, I'd be remiss if I didn't offer a special salute to Dr. Keith Notley, Kingston, Ontario, whose 1980 doctoral thesis investigated the cause of the Springhill mine disaster. Keith was my authoritative source on mines and mining methodology. And a tip of my lens cap to photographer Walter ("Wally") Hayes and his wife, Joyce, Halifax. I believe Wally is the last surviving member of the media contingent who rushed to Springhill to report on the Bump. Years later, he served as the official photographer of the Province of Nova Scotia (1986–2011). Not only was Wally kind enough to share with me his unpublished memoir of his Springhill experiences, but he also introduced me to many knowledgeable and wonderfully helpful people in Springhill, and he went above and beyond in providing me with some of his superb photos, several of which appear in the pages of this book.

I'd also like to thank the following individuals for their help as I went about the research and writing of *Blood on the Coal*: Roberta Bell, Springhill, NS; Elizabeth (Ryan) Blackwell, Brandon,

FL; Marjorie Bousfield, Wolfe Island, ON; Doris Brotman and her son Judd, Wellesley, MA; Dr. Peter Calder (the son of Springhill mine manager George Calder), Kingston, ON; Colin Cavada, communications assistant, Carnegie Hero Fund Commission, Pittsburgh, PA; Bob Clark, Springhill, NS; Verna (Allison) Clark, Springhill, NS; historian-author Pat Crowe, Springhill, NS; actor Beau Dixon (creator of the play *Beneath Springhill: The Maurice Ruddick Story*), Peterborough, ON; Russell Fisher, Springhill, NS; Jacqueline Foster, senior communications advisor, Nova Scotia Power, Halifax, NS; historian Alice M. Gibb, London, ON; Melissa Fay Greene, author of the 2003 book *Last Man Out: The Story of the Springhill Mine Disaster*, Atlanta, GA; Amos Grogan, Springhill, NS; my "go-to" graphic designer, Larry Harris, Kingston, ON; Noah Harrison, tour guide, Springhill Miners' Museum, Springhill, NS; Janice Hill, associate vice-principal (Indigenous Initiatives and Reconciliation), Queen's University, Kingston, ON; Julia Holland, archives technician, Clara Thomas Archives and Special Collections, York University Library, Toronto ON; Eleanor and Daniel Jolly, Amherst, NS; Billy Kempt, Springhill, NS; Jessica Kilford, Nova Scotia Archives, Halifax, NS; the staff of Library and Archives Canada; Mary Linnemann, digital imaging coordinator, Hargrett Rare Book and Manuscript Library, University of Georgia, Athens, GA; archives researcher Kristen Lipscombe, Halifax, NS; mining historian-photographer Mary Willa Littler, Springhill, NS; Darryl MacKenzie, curator, and the staff of the Cumberland County Museum and Archives, Amherst, NS; Anna MacNeil, archival research assistant, Beaton Institute, Cape Breton University, Sydney, NS; Roberta McMaster, manager, Springhill Miners' Museum, Springhill, NS; archives researcher Maureen McNeil, Cape Breton, NS; award-winning food writer and culinary historian Lindy Mechefske, Kingston, ON; Lorraine Munroe, Amherst, NS; Barbara Murphy, permissions coordinator, National Academies Press, Washington, DC; Anne Murray, Canada's first lady of song, Halifax, NS; Dean Ruddick and his partner, Joyce Capon, Springhill, NS; Audrey (Skidmore) Ryan, Amherst, NS; Murray Scott, mayor of the Municipality of Cumberland, Springhill, NS; Gordon E. Smith, DAN School of Drama and Music, Queen's University, Kingston, ON; Fencebusters' baseball historian Doug Spencer, Springhill, NS; archives researcher Taylor Starr, Maple, ON; Denise Taylor, Museum of Industry, Stellarton, NS; Mary Louise Wilder-Smith, Fredericton, NB; and Erika Wilson, Museum of Industry, Stellarton, NS.

And last, but certainly not least, I owe a special thank you to my wife, Marianne Hunter, the love of my life, for her encouragement and patience and for being so understanding when I missed family events and other social gatherings so I could "work on the book."

The author (*left*) with Harold Brine (*right*) in 2022. (Courtesy of Murriel Brine)

Notes

EPIGRAPH

Bono, "The Resource Miracle," *TIME*, May 28, 2012, https://content
.time.com/time/magazine/article/0,9171,2115044-1,00.html.

INTRODUCTION

1. SaltWire Network, October 22, 2018, www.saltwire.com/nova-scotia
   /news/did-springhill-miners-see-the-bump-coming-a-look-at-the-science
   -behind-the-1958-mining-disaster-252565.
2. That wonderfully evocative phrase appears in the 1959 tune "The Ballad
   of Springhill," which was written by American folksinger Peggy Seeger
   and her English partner, Ewan MacColl. Over the years, that ditty has
   become the most famous and most often performed Springhill mine
   disaster song. For more information, please visit https://disastersongs.ca
   /the-ballad-of-springhill.
3. *Globe and Mail*, November 19, 1958, p. 6.
4. Author interviews/conversations with miner Harold Brine, during the period of August 7, 2021, to December 20, 2022. (Cited hereafter as Harold
   Brine interviews.)
5. *Globe and Mail*, October 31, 1958, p. 6.
6. Printed with permission of Valerie (Ruddick) MacDonald.
7. https://news.stanford.edu/2005/06/12/youve-got-find-love-jobs-says.
8. McKay, Cheryl, *Spirit of Springhill: Miners, Wives, Widows, Rescuers
   & Their Children Tell True Stories of Springhill's Coal Mining Disasters*
   (Atlanta: Purple PenWorks, 2014, Kindle edition), p. 44.
9. Eliot, George, *Daniel Deronda* (London: William Blackwood and Sons,
   1876), chapter 5.

CHAPTER 1

1. Scott, Bertha Isabel, *Springhill: A Hilltop in Cumberland* (Springhill, NS: Privately printed, 1926), p. 9.
2. It was a German meteorologist named Alfred Wegener (1880–1930) who coined the word *Pangea*—which is an invented creation with Greek, German, and English roots that means "all lands." It was also Wegener who, in 1912, devised the first comprehensive theory of continental drift—the idea that Earth's continents slowly move relative to one another.
3. Frame, Elizabeth, *A List of Micmac Names of Places, Rivers, etc. in Nova Scotia* (London: John Wilson and Sons University Press, 1892), p. 9.
4. British soldier Lieutenant Colonel Robert Monckton applied the name "Cumberland" to Fort Beauséjour, New Brunswick, which was captured from the French in June of 1755. Monckton did so to honour William Augustus, the Duke of Cumberland (the third son of King George II), who had commanded the British forces that routed Scottish Jacobite rebels at the Battle of Culloden in 1746.
5. Freese, Barbara, *Coal: A Human History* (New York: Perseus Publishing, 2003), p. 2.
6. As quoted by John Calder in his booklet *Coal in Nova Scotia* (Halifax: Nova Scotia Department of Mines and Energy, 1985), p. 9.
7. "Not Your Grandfather's Mining Industry," home page of the Nova Scotia Mining Association, https://notyourgrandfathersmining.ca/general-mining-association.
8. Chignecto Coal Mining Company prospectus, Amherst, Nova Scotia (1864). Online at https://catalog.hathitrust.org/Record/100311855/Cite.
9. Campbell, Bertha J., *Springhill: Our Goodly Heritage* (Springhill, NS: Springhill Heritage Group, 1989), p. iii.
10. Gesner accompanied the Scottish-born geologist Sir Charles Lyell—the father of modern geology—on his 1842 visit to Joggins, after which he declared that the "forest of fossil coal-trees" exposed in the seaside cliffs was "the most wonderful phenomenon perhaps that I have seen." It was Lyell's examination of this fossil forest that settled the scientific debate about the origins of coal. He confirmed that coal was decayed plant matter from ancient forests.
11. Campbell, *Springhill: Our Goodly Heritage*, p. iii.

CHAPTER 2

1. Harold Brine interviews.
2. Stewart, Greig, *Arrow Through the Heart: The Life and Times of Crawford Gordon and the Avro Arrow* (Toronto: McGraw-Hill Ryerson, 1998), p. 2.

3. *Globe and Mail*, August 29, 1957, p. 22.
4. *Globe and Mail*, October 29, 1957, p. 22.
5. Author interview with Billy Kempt, January 4, 2022. (Cited hereafter as Billy Kempt interview.)
6. Notley, Keith, "Analysis of the Springhill Mine Disaster" (Kingston, ON: doctoral thesis written for the Mining Engineering Department, Queen's University, ON, 1980), p. 153.
7. Notley, "Analysis of the Springhill Mine Disaster," p. 3.
8. SaltWire Network, October 22, 2018, www.saltwire.com/nova-scotia /news/did-springhill-miners-see-the-bump-coming-a-look-at-the-science -behind-the-1958-mining-disaster-252565.

CHAPTER 3

1. For the full story, please visit https://notyourgrandfathersmining.ca /mother-coo.
2. *Ottawa Citizen*, October 24, 1958, p. 7.
3. Blank, Joseph, "The Big 'Bump' at Springhill," *Reader's Digest*, January 1960, pp. 97–98.
4. From a December 12, 1930, article written by David P. Sentner, a correspondent for the International News Service who interviewed Einstein.
5. Harold Brine interviews (all quotations in this section).
6. Harold Brine interviews (all quotations in this section unless otherwise noted).
7. Article quoted on the web site https://notyourgrandfathersmining.ca /child-labour.
8. According to a 2010 article, "The Dangers of Mining Around the World," which appears on the BBC website. www.bbc.com/news/world-latin -america-11533349.
9. Harold Brine interviews (all quotations in this section).

CHAPTER 4

1. Author interview with Verna (Allison) Clark, Springhill, NS, May 10, 2022.
2. Crowe, Pat, "The Heritage Corner," *Springhill Record*, November 16, 2005.
3. Springhill historian Pat Crowe chronicled the history of the Liars Bench in the December 14, 2005, edition of her "Heritage Corner" column, which appeared in the *Springhill Record*. She wrote, "This bench was put together from scrap lumber by some of the older men of the town, such as Percy (Pa) Tabor, Danny Ross, Hubert Guthro, and Bill Brown."
4. Greene, Melissa Fay, *Last Man Out: The Story of the Springhill Mine Disaster* (New York: Harcourt, 2003), p. 25.

5. Tabor family interview with Melissa Fay Greene, July 1997. (University of Georgia Archives.)

CHAPTER 5

1. Author interview with Valerie (Ruddick) MacDonald, November 24, 2021. (Cited hereafter as Valerie [Ruddick] MacDonald interview.)
2. Maurice Ruddick interview with Dr. Noel Murphy, 1958, www.cbc.ca /listen/live-radio/1-27-information-morning-ns/clip/15900471-archival -recording-maurice-ruddick-speaking-springhill-mining-disaster.
3. Norma Ruddick interview with Melissa Fay Greene, July 6, 1997. (Cited hereafter as Norma Ruddick interview.)
4. Norma Ruddick interview.
5. Valerie (Ruddick) MacDonald interview.
6. *Toronto Telegram*, November 6, 1958, p. 6.
7. Norma Ruddick interview.
8. Valerie (Ruddick) MacDonald interview.
9. Thomas, Verna, *Invisible Shadows* (Halifax: Nimbus Publishing, 2001), pp. 58–59.
10. Norma Ruddick interview.
11. Author interview with Anne Murray, September 10, 2021. (Cited hereafter as Anne Murray interview.)
12. Harold Brine interviews.
13. Boutilier, Danny, "When You Work Down in the Old Number Two." Available online at https://disastersongs.ca/down-in-springhills-bumpy -mine.
14. McKay, Ian, "The Realm of Uncertainty: The Experiences of Work in the Cumberland Coal Mines, 1873–1927," *Acadiensis*, Autumn 1986, Vol. 16, No. 1, p. 23.
15. Valerie (Ruddick) MacDonald interview.

CHAPTER 6

1. Burden, Dr. Arnold, *The Dramatic Life of a Country Doctor: Fifty Years of Disasters and Diagnoses* (Halifax: Nimbus Publishing, 2011), p. 88.
2. Burden, *Dramatic Life of a Country Doctor*, p. 95.
3. Author interview with Amos Grogan, January 24, 2022.
4. Author interview with Bill Burden, November 2, 2021. (Cited hereafter as Bill Burden interview.)
5. Burden, *Dramatic Life of a Country Doctor*, p. 102.
6. Murray, Anne, *All of Me* (Toronto: Knopf Canada, 2010, Kindle edition), p. 9. (Cited hereafter as *All of Me*.)
7. Murray, *All of Me*, p. 7.

8. Carson Murray died in 1980 at age seventy-two. He'd suffered from lymphoma for several years, an affliction that his daughter, Anne, suspects was the result of years of exposure to X-ray radiation at the hospital. His liver had also been damaged by the after-effects of an old case of hepatitis C that he likely contracted in the operating room. Marion Murray passed in 2006 at the age of ninety-two as result of complications from heart surgery.
9. Author interview with Russell Fisher, March 17, 2022.
10. Burden, *Dramatic Life of a Country Doctor*, p. 21.
11. Burden, *Dramatic Life of a Country Doctor*, p. 74.
12. Burden, *Dramatic Life of a Country Doctor*, p. 78.

CHAPTER 7

1. Reynolds managed to solicit this "finding" by providing doctors a complimentary carton of Camels and then asking what brand they smoked. As the *New York Times* reported on October 6, 2008, "From the 1920s into the 1950s, cigarette ads featured endorsers as varied as babies, Mickey Mantle, doctors and even Santa Claus." www.nytimes.com/2008/10/07/business/media/07adco.html.
2. Herb Pepperdine interview with Melissa Fay Greene, April 8, 1998 (University of Georgia Archives). (Cited hereafter as Herb Pepperdine interview.)
3. Those are the latest numbers compiled by the United States Department of Labor.
4. *The Globe* (Toronto), August 11, 1879, p. 2.
5. Iowa-born John L. Lewis (1880–1969), the largely self-educated, Shakespeare-quoting labour leader, headed the United Mine Workers of America for forty years, from 1920 until 1960.
6. Harold Brine interviews.
7. An Edison battery, which was 6 inches wide by 8 inches long and 1.5 inches thick, weighed about 2.5 pounds. A flexible cord ran along the helmet's crown. A guide at the back of the helmet routed the cable to the battery pack on the miner's belt. The lamps were about 75 lumens, which is the equivalent of 15 watts.
8. His full name was Clarence Percy Rector, but he went by his middle name.
9. Herb Pepperdine interview.
10. Munroe, Harry, *Brave Men of the Deep: A Part of the History of Springhill's Mining* (Bloomington, IN: Author House, 2014), p. 29.
11. Lyrics printed with permission of Valerie (Ruddick) MacDonald.
12. Burden, *Dramatic Life of a Country Doctor*, p. 74.
13. Blank, "The Big 'Bump' at Springhill," p. 98.

14. Norma Ruddick interview.
15. Herb Pepperdine interview.

CHAPTER 8
1. *Science Daily*, July 6, 2020, www.sciencedaily.com/releases/2020/07/200706101837.htm.
2. Frost testimony, Proceedings of the Province of Nova Scotia Public Inquiry into the Springhill Mine Disaster, January 1959, p. 57. (Cited hereafter as Springhill Mine Disaster Inquiry.)
3. Harold Brine interviews.
4. Bowman Maddison interview with Dr. Noel Murphy, Dr. H.D. Beach, and Dr. R.A. Lucas, circa 1959. (Cited hereafter as Bowman Maddison interview.)
5. *Maclean's*, October 29, 1988.

CHAPTER 9
1. The noxious gases that plague mines are referred to as "damps"—a term that has its roots in the German word *dampf*, which translates into English as "vapour."
2. Harold Brine interviews.
3. Greene, *Last Man Out*, p. 30.
4. Bowman Maddison interview.
5. Bowman Maddison interview.

CHAPTER 10
1. Author interview with Peter Calder, son of George, September 29, 2021. (Cited hereafter as Peter Calder interview.)
2. Brown, James, *Miracle Town: Springhill, N.S.* (Hantsport, NS: Lancelot Press, 1983), p. 90.
3. Springhill Mine Disaster Inquiry, p. 88.
4. Burden, *Dramatic Life of a Country Doctor*, p. 144.
5. Burden, *Dramatic Life of a Country Doctor*, p. 144.
6. Anne Murray interview.
7. Billy Kempt interview.
8. Billy Kempt interview.
9. Valerie (Ruddick) MacDonald interview.

CHAPTER 11
1. Harold Brine interviews.
2. Springhill Mine Disaster Inquiry, p. 48.
3. Springhill Mine Disaster Inquiry, p. 53.

4. Bev Reynolds, 1993 "Women of Springhill" interview with Marjorie Whitelaw, CBC Radio's *Maritime Magazine*, broadcast transcript, p. 3 (University of Georgia archives). Reynolds, who worked at the Springhill Medical Centre, had feared for the life of her miner husband when the Bump occurred. Sadly, as it turned out, Frederick Wesley Reynolds was among the seventy-five men who died. As his death certificate reported, he died instantly owing to "most severe contusions and crushing."
5. *Halifax Chronicle-Herald*, October 25, 1958, p. 2.
6. *Halifax Chronicle-Herald*, October 25, 1958, p. 9.
7. *Halifax Chronicle-Herald*, October 25, 1958, p. 9.
8. Death certificate completed by Dr. D. Fisher, November 10, 1958. Available online at https://archives.novascotia.ca/vital-statistics/death/?ID=414468.
9. It was a German scientist named Alexander Bernhard Dräger who in 1928 invented a combination gas mask and oxygen inhalator as a breathing apparatus for underground rescue workers. According to the Nova Scotia Museum of Industry home page: "As early as 1870, breathing equipment existed and was used in British mines. In 1906–07 the Dominion Coal Company set up a rescue station in Glace Bay with 20 sets of breathing apparatus, related equipment, and first aid supplies, the first such facility in North America. The equipment brand was Draeger—thus 'draegermen' came to be the name for mine rescue workers. (The *Oxford Dictionary* attributes the name as Canadian so it appears that the name was coined in Nova Scotia with those first rescue workers.)" For more information, please visit https://museumofindustry.novascotia.ca.
10. *Halifax Chronicle-Herald*, October 25, 1958, p. 9.
11. Mabel and Currie Smith interview with Melissa Fay Greene, April 8, 1998, p. 3.
12. *Globe and Mail*, September 11, 2018.

CHAPTER 12

1. *Springhill-Parrsboro Record*, October 30, 1958, p. 3.
2. Beach, Horace D., and Lucas, Rex A., Disaster Study Number 13: *Individual and Group Behavior in a Coal Mine Disaster* (Washington, DC: National Academy of Sciences–National Research Council, 1960), p.18. Available online at www.ilankelman.org/miscellany/DS13Beach.pdf.
3. In 1958, the Herald Limited published two newspapers—the morning *Chronicle-Herald*, which was distributed province-wide, and the afternoon *Mail-Star*, which was read by people living in the Halifax–Dartmouth area.

4. *Halifax Chronicle-Herald*, October 24, 1958, p. 1.
5. Hayes, Walter, *Recollections of the Springhill Mine Disaster of 1958* (unpublished memoir, 2008).
6. Hayes, *Recollections of the Springhill Mine Disaster*.
7. In addition to Casey, the other members of the *Chronicle-Herald* news team in Springhill included reporters Gordon Duffy (who was in charge) and Ron Slade, as well as photographers Bob Norwood, Harry Cochrane, Maurice Slaunwhite, Ben McCall, and Wally Hayes.

CHAPTER 13

1. Harold Brine interviews.
2. Harold Brine interviews.
3. Blank, "The Big 'Bump' at Springhill," p. 98.
4. Miner Harold Brine cautioned the author of this book that he has spotty recollections of what was said in the days when he and his companions were trapped underground. "I only know what I said and did. I can't tell you what the other boys were saying or doing," he advised. This bit of dialogue between Levi Milley and Caleb Rushton, as well as other exchanges throughout the book in the "created" narrative between the men who were trapped underground, is included to add vibrancy and give the reader a sense of time and place; however, there is one important caveat: The words are rooted in a variety of sources that hopefully are reliable. They include myriad contemporary newspaper reports, the transcripts of the many interviews conducted by various researchers over the years, and the dialogue that's offered up in two earlier accounts of the disaster. One is *Miracle at Springhill*, the 1960 book by *Boston Globe* reporter Leonard Lerner. The other is *Last Man Out*, the 2003 book by Georgia writer Melissa Fay Greene. Both authors crafted detailed dialogues that supposedly were engaged in by the men trapped in the shattered No. 2 mine. Lerner wrote his book a year after the disaster, and it's likely that he spoke to many of the miners personally. That being so, the men probably still had a fresh sense of the gist of what was said and when during their torturous ordeal. But we still cannot expect anyone to recount the exact words of those living through an event as traumatic as a mine disaster. Greene interviewed some of the surviving miners forty years *after* the disaster, and who can say for certain how accurate their memories were of the conversations and events that unfolded in the pitch-black mine? The Greene book, which includes many pages of dialogue, stands as the definitive source for what happened underground and who among the trapped miners said what and when. That said, both the Lerner and Greene books read as though the narrators were there to observe events and record the conversations that are recounted. In the interests of full disclosure, it's only

fair to caution the reader that the narrative in this book utilizes the same literary techniques that Lerner and Greene used, albeit to a limited extent and to a bare minimum; definitive sources are otherwise noted.

5. Joe McDonald interview with Melissa Fay Greene, 1997.
6. Bowman Maddison interview.
7. Larry Leadbetter interview with Melissa Fay Greene, 1998. (Cited hereafter as Leadbetter interview.)
8. Harold Brine interviews.

CHAPTER 14

1. CTV interview, October 23, 2008, www.ctvnews.ca/springhill-n-s-marks -50-years-since-mining-disaster-1.336403.
2. American Battlefield Trust, www.battlefields.org/learn/articles/amputations -and-civil-war.
3. *Springhill-Parrsboro Record*, November 6, 1958, p. 1.

CHAPTER 15

1. *Halifax Chronicle-Herald*, October 24, 1958, p. 1.
2. *Halifax Chronicle-Herald*, October 24, 1958, p. 1.
3. Lerner, Leonard, *Miracle at Springhill* (New York: Holt, Rinehart and Winston, 1960), p. 26.
4. Springhill Mine Disaster Inquiry, p. 90.
5. Over the next two weeks of effort, a total of fifty-six draegermen were involved in rescue missions in the No. 2 mine. Ten of these men were Springhillers: Wilfred Brown, Amos Cogan, Arthur Dobson, Victor Hunter, William James, Garnet Jewkes, Hilton MacNutt, Kenneth Megeney, Harry Munroe, and Neil Ross.
6. Burden, *Dramatic Life of a Country Doctor*, p. 145.
7. Burden, *Dramatic Life of a Country Doctor*, p. 145.
8. Burden, *Dramatic Life of a Country Doctor*, p. 147.
9. Burden, *Dramatic Life of a Country Doctor*, p. 147.
10. Burden, *Dramatic Life of a Country Doctor*, p. 148.
11. Burden, *Dramatic Life of a Country Doctor*, p. 150.
12. Burden, *Dramatic Life of a Country Doctor*, p. 151.
13. Anne Murray interview.

CHAPTER 16

1. *Springhill-Parrsboro Record*, October 30, 1958, p. 3.
2. *Springhill-Parrsboro Record*, October 30, 1958, p. 5.
3. Lerner, *Miracle at Springhill*, p. 61.
4. Lerner, *Miracle at Springhill*, p. 61.
5. Lerner, *Miracle at Springhill*, p. 62.

6. *Halifax Chronicle-Herald* (extra edition), October 27, 1958, p. 2.
7. For more information, please visit www.cbc.ca/player/play/1834738154.
8. Willis's reports were carried by the BBC and broadcast to Europe, and they were heard on 650 radio stations in the United States, including affiliates of the NBC network. For more information, please visit www .cbc.ca/player/play/1834738154.
9. For more information, please visit https://broadcasting-history.com/listing _and_histories/television/cbht-dt.
10. Interview available online at cbc.ca/player/play/2459565984.
11. To view video of the 1958 coverage of the disaster, please visit www.cbc .ca/player/play/1826341159.
12. Blank, "The Big 'Bump' at Springhill," p. 107.
13. One of the more bizarre—and jarring—stories of the mine disaster came to light when Mayor Gilroy told a *Globe and Mail* reporter (October 24, 1958, p. 2) that among the flood of letters that had arrived in his mailbox were "at least twenty from lonely bachelors who wanted to marry women widowed by Springhill's most recent misfortune."
14. *Globe and Mail*, October 25, 1958, p. 2.
15. Billy Kempt interview.

CHAPTER 17

1. Roy, Andrew, *History of the Coal Miners of the United States* (Columbus, OH: J.L. Trauger Printing, 1907), p. 376.
2. Harold Brine interviews.

CHAPTER 18

1. "The Seven" was a term the media coined as shorthand for the group of miners who found themselves trapped at the top end of the 13,000-foot coal face. There actually were eight men in this group, but for reasons explained in the course of this narrative, only seven men were referenced.
2. Maurice Ruddick interview with Drs. Murphy, Beach, and Lucas, 1959, p. 9.
3. Harold Brine interviews.

CHAPTER 19

1. *Halifax Chronicle-Herald*, October 27, 1958, p. 6.
2. Hayes, *Recollections of the Springhill Mine Disaster*.
3. Hayes, *Recollections of the Springhill Mine Disaster*.
4. Hayes, *Recollections of the Springhill Mine Disaster*.
5. Hayes, *Recollections of the Springhill Mine Disaster*.
6. *Halifax Chronicle-Herald*, October 27, 1958, p. 6.
7. Hayes, *Recollections of the Springhill Mine Disaster*.

8. Greene, *Last Man Out*, p. 136.
9. Author interview with Valarie (Tabor) Alderson, June 9, 2022. (Cited hereafter as Valarie [Tabor] Alderson interview 2.)
10. *Toronto Telegram*, November 12, 1958, p. 6.
11. Harold Brine interviews.

CHAPTER 20
1. Valarie (Tabor) Alderson interview 2.
2. Valarie (Tabor) Alderson interview 2.
3. *Boston Globe*, October 26, 1958, p. 1.
4. Lerner, *Miracle at Springhill*, p. 92.
5. *Globe and Mail*, October 31, 1958, p. 1.
6. Hayes, *Recollections of the Springhill Mine Disaster*.

CHAPTER 21
1. Lerner, *Miracle at Springhill*, p. 109.
2. Greene, *Last Man Out*, pp. 104–105.
3. Lyrics printed with permission of Valerie (Ruddick) MacDonald.
4. Frank Hunter interview with Drs. Murphy, Beach, and Lucas, 1958, p. 28.
5. Herb Pepperdine interview with Drs. Murphy, Beach, and Lucas, 1959.

CHAPTER 22
1. Hayes, *Recollections of the Springhill Mine Disaster*.
2. Hayes, *Recollections of the Springhill Mine Disaster*.
3. Harold Brine interview with Drs. Murphy, Beach, and Lucas, 1958, p. 81.

CHAPTER 23
1. Harold Brine interview with Drs. Murphy, Beach, and Lucas, 1958, p. 81.
2. Harold Brine interview with Drs. Murphy, Beach, and Lucas, 1958, p. 85.
3. Hayes, *Recollections of the Springhill Mine Disaster*.
4. Hayes, *Recollections of the Springhill Mine Disaster*.

CHAPTER 24
1. Blank, "The Big 'Bump' at Springhill," p. 106.
2. Burden, *Dramatic Life of a Country Doctor*, p. 160.
3. Burden, *Dramatic Life of a Country Doctor*, p. 160.
4. A video of the CBC television coverage of the Springhill mine rescue on the evening of October 29, 1958, can be found at www.cbc.ca/player/play /1826341159.
5. Burden, *Dramatic Life of a Country Doctor*, p. 162.
6. Lerner, *Miracle at Springhill*, p. 151.

7. *Halifax Chronicle-Herald*, October 31, 1958, p. 7.
8. *Canada: This Land, These People* (Montreal: Reader's Digest Association, 1968), p. 106.
9. Burden, *Dramatic Life of a Country Doctor*, p. 162.

CHAPTER 25

1. *Globe and Mail*, October 31, 1958, p. 6.
2. *Halifax Chronicle-Herald*, November 3, 1958, p. 4.
3. Relief funds for the Springhill miners and their dependants were also started in Toronto and in Boston, Massachusetts.
4. Greene, *Last Man Out*, p. 295.
5. *Halifax Chronicle-Herald*, October 31, 1958, p. 7.
6. *Halifax Chronicle-Herald*, October 31, 1958, p. 7.
7. *Globe and Mail*, October 31, 1958, p. 1.
8. *Halifax Chronicle-Herald*, October 31, 1958, p. 3.
9. *Halifax Chronicle-Herald*, October 31, 1958, p. 3.
10. *Halifax Chronicle-Herald*, November 1, 1958, p. 1.
11. *Toronto Telegram*, November 1, 1959, p. 11-1.
12. Lerner, *Miracle at Springhill*, p. 162.
13. *Toronto Daily Star*, November 1, 1958, p. 9.
14. *Halifax Chronicle-Herald*, November 1, 1958, p. 1.
15. For the full story on the comedian's appearance on *The Ed Sullivan Show*, see Greene, *Last Man Out*, p. 126.
16. Greene, *Last Man Out*, p. 126.

CHAPTER 26

1. Burden, *Dramatic Life of a Country Doctor*, p. 164.
2. Burden, *Dramatic Life of a Country Doctor*, p. 166.
3. Burden, *Dramatic Life of a Country Doctor*, p. 166.
4. Author interview with Wally Hayes, November 21, 2021. (Cited hereafter as Wally Hayes interview.)
5. *Halifax Chronicle-Herald*, November 3, 1958, p. 3.
6. Lyrics printed with permission of Valerie (Ruddick) MacDonald.
7. *Toronto Telegram*, November 3, 1958, p. 3.
8. *Toronto Telegram*, November 3, 1958, p. 1.
9. Burden, *Dramatic Life of a Country Doctor*, p. 168.
10. Stewart, *Arrow Through the Heart*, p. 148.

CHAPTER 27

1. *Globe and Mail*, November 13, 1958, p. 1.
2. *Toronto Telegram*, November 3, 1958, p. 1.
3. Wally Hayes interview.

4. Author interview with Valarie (Tabor) Alderson, December 7, 2021. (Cited hereafter as Valarie [Tabor] Alderson interview 1.)
5. *Toronto Daily Star*, November 3, 1958, p. 1.
6. *Toronto Daily Star*, November 3, 1958, p. 1.
7. *Toronto Telegram*, November 3, 1958, p. 5.
8. Harold Brine interviews.
9. *Toronto Telegram*, November 2, 1958, p. 6.
10. *Toronto Telegram*, November 3, 1958, p. 5.
11. *Toronto Telegram*, November 3, 1958, p. 5.
12. *Life*, December 8, 1958, p. 49. To view *Life* magazine's coverage of the trip, please visit the Google books website.
13. Greene, *Last Man Out*, p. 270.
14. *Globe and Mail*, November 24, 1958, p. 21.
15. *Globe and Mail*, November 24, 1958, p. 21.
16. *Globe and Mail*, November 24, 1958, p. 21.

CHAPTER 28

1. Gilroy would stand as the Liberal candidate in the Cumberland Centre riding in the June 1960 provincial election, coming in second to the Conservative candidate. Ultimately, the strain of that campaign and his work as mayor of Springhill adversely affected his health. After falling ill in early 1961, he spent three months in hospital. Then, in August of that year, he died at the age of fifty-one.
2. Herb Pepperdine interview.
3. *Globe and Mail*, November 13, 1958, p. 1.
4. *Globe and Mail*, October 23, 1959, p. 9.
5. *The Empire Club of Canada Speeches* 1953–54 (Toronto: The Empire Club Foundation, 1954), pp. 29–40.
6. *Halifax Chronicle-Herald*, January 29, 1959, p. 1.
7. To access the report, please visit www.mininghistory.ns.ca/index2 .htm#spring.
8. McInnes Commission Report, p. 6.
9. Keith Notley's calculations can be found in his 1980 doctoral thesis, "Analysis of the Springhill Mine Disaster" (written for the Mining Engineering department at Queen's University in Kingston, Ontario).
10. This provocative novel deals with the disastrous efforts by a group of British schoolboys who are stranded on an uninhabited island to govern themselves. Among the themes Golding dealt with are the tensions between groupthink and individuality, between rational and emotional reactions, and between morality and immorality.
11. *On the Beach*, an apocalyptic novel written by British author Nevil Shute, tells the story of a group of people in Melbourne, Australia, who are

awaiting the arrival of a cloud of deadly radiation that's drifting toward them from the Northern Hemisphere in the wake of a nuclear war.

12. Beach and Lucas, Disaster Study Number 13, p. 140.
13. Beach and Lucas, Disaster Study Number 13, p. 140.
14. *Globe and Mail*, November 7, 1958, p. 1.
15. *Globe and Mail*, May 26, 1959, p. 1.
16. *Toronto Telegram*, January 22, 1959, p. 1.
17. Greene, *Last Man Out*, p. 288.
18. Peter Calder interview.

EPILOGUE

1. For more information, please visit www.geothermal-energy.org/pdf /IGAstandard/WGC/1995/1-jessop.pdf.
2. Valarie (Tabor) Alderson interview 1.
3. *Canadian Panorama*, May 10, 1969.
4. To view the Historica Canada video about Maurice Ruddick (which, incidentally, erroneously reports that he'd suffered a broken leg in the Bump), please visit www.historicacanada.ca/content/heritage-minutes /maurice-ruddick.
5. Author interview with Beau Dixon, March 14, 2022.
6. The controversial Donkin mine, the only underground coal mine still operating in Nova Scotia, is also the world's only operating subsea coal mine. It faces an uncertain future at best.
7. https://novascotia.ca/natr/meb/geoscience-online/about-database-amo.asp.
8. According to numbers that are posted on the website of the Nova Scotia Museum of Industry: https://museumofindustry.novascotia.ca/coal-mining -fatalities.
9. *Globe and Mail*, September 26, 2022, p. A10.
10. www.worldcoal.org/coal-facts.
11. www.nsenergybusiness.com.
12. For more information, please visit www.facebook.com/SpringhillMiners Museum.
13. Author interview with Roberta Bell, January 18, 2022.
14. To view the Blood on the Coal tartan, please visit the Scottish Register of Tartans home page at www.tartanregister.gov.uk/tartanDetails?ref=12677.

ACKNOWLEDGEMENTS

1. https://journal.neilgaiman.com/2004/05/pens-rules-finishing-things-and -why.asp?fbclid=IwAR37InspsmVlFaqdtyTeBCBU4KTBqWGceJggN -EGYBvgSpEYSEEbvTot_so.

SELECT BIBLIOGRAPHY

BOOKS

Beach, Horace D., and Lucas, Rex A., Disaster Study Number 13: *Individual and Group Behavior in a Coal Mine Disaster* (Washington, DC: National Academy of Sciences–National Research Council, 1960).

Brown, James B., *Miracle Town: Springhill, N.S.* (Windsor, NS: Lancelot Press, 1983).

Brown, Roger D., *Blood on the Coal: The Story of the Springhill Mining Disasters* (Halifax: Nimbus Publishing, revised edition, 2002).

Burden, Dr. R. Arnold, *The Dramatic Life of a Country Doctor: Fifty Years of Disaster and Diagnoses* (Halifax: Nimbus Publishing, 2011).

Calder, John, *Coal in Nova Scotia* (Halifax: Nova Scotia Department of Mines and Energy, 1985).

Campbell, Bertha J., *Springhill: Our Goodly Heritage* (Springhill, NS: Springhill Heritage Group, 1989).

Frame, Elizabeth, *A List of Micmac Names of Places, Rivers, etc. in Nova Scotia* (London: John Wilson and Sons University Press, 1892).

Greene, Melissa Fay, *Last Man Out: The Story of the Springhill Mine Disaster* (New York: Harcourt, 2003).

Hayes, Walter, *Recollections of the Springhill Mine Disaster of 1958* (unpublished memoir, 2008).

Lerner, Leonard, *Miracle at Springhill* (New York: Holt, Rinehart and Winston, 1960).

McKay, Cheryl, *Spirit of Springhill: Miners, Wives, Widows, Rescuers & Their Children Tell True Stories of Springhill's Coal Mining Disasters* (Atlanta: Purple PenWorks, 2014).

Muise, Delphin A., and McIntosh, Robert G., *Coal Mining in Canada: A Historical and Comparative Overview* (Ottawa: National Museum of Science and Technology, 1996).

Munroe, Harry E., *Brave Men of the Deep: A Part of the History of Springhill's Mining* (Bloomington, IN: Author House, 2014).

Notley, Keith, "Analysis of the Springhill Mine Disaster" (Kingston, ON: doctoral thesis written for the Mining Engineering Department, Queen's University, ON, 1980).

Roy, Andrew, *History of the Coal Miners* (Columbus, OH: J.L. Trauger Printing, 1907).

Sandlos, John, and Keeling, Arn, *Mining Country: A History of Canada's Mines and Miners* (Toronto: James Lorimer & Company, 2021).

Scott, Bertha Isabel, *Springhill: A Hilltop in Cumberland* (Springhill, NS: privately published, 1926).

NEWSPAPER AND PERIODICAL ARTICLES

Anglican Journal, October 2008, www.anglicanjournal.com.

Blank, Joseph P., "The Big 'Bump' at Springhill," *Reader's Digest*, January 1960.

Boston Globe, October 24–November 1, 1958; November 3, 1958, p. 9; November 6, 1958, p. 9; November 7, 1958, p. 45; November 14, 1958, p. 28; November 24, 1958, p. 17; November 29, 1958, p. 63; October 2, 1960, p. 61; October 31, 1978, p. 2.

Canadian Panorama, May 10, 1969.

DeMont, John, "The other disaster," *Maclean's*, October 28, 2002, p. 64.

Finlayson, Ann, "'Blood on the coal,'" *Maclean's*, October 24, 1988, pp. N1–3.

Global Energy Monitor, www.gem.wiki/Coal_mining_disasters#Canada.

Globe and Mail, October 24, 1958, p. 1; October 25, 1958, pp. 1, 2; October 27, 1958, p. 1; November 1, 1958, pp. 1, 2, 7; November 4, 1958, p. 1; November 7, 1958, p. 2; November 19, 1958, p. 6; December 12, 1958, p. 9; December 27, 1958, pp. A-1, 13, 24; December 31, 1958, p. 3; January 28, 1959, p. 11; January 30, 1959, p. 9; January 4, 1960, p. 23; April 29, 2023, pp. A10–12.

Halifax Chronicle-Herald, October 24–November 6, 1958.

Harrison, Hope, "The life and death of the Cumberland mines," *Nova Scotia Historical Review* 5, No. 1 (1985), pp. 75–83.

Life, November 3, 1958, p. 39; November 10, 1959, pp. 23–27.

New York Times, October 24, 1958, p. 1; November 2, 1958, p. 1.

Ottawa Citizen, October 24, 1958, p. 7.

Springhill-Parrsboro Record, October 23, October 30, November 6, November 13, November 20, November 27, December 4, 1958.

Toronto Star, October 24–November 3, 1958.

Toronto Telegram, October 24–November 2, 1958; November 3, 1958, pp. 1, 3.

ONLINE

Beaton Institute, Cape Breton University, www.cbu.ca/community/beaton
-institute.

Canadian Encyclopedia, www.thecanadianencyclopedia.ca/en/article/coal.

CBC Archives, www.cbc.ca/archives/history.

CBC News, "Cape Breton's Donkin coal mine fined for safety violations after
fall reopening," January 19, 2023, www.cbc.ca/news/canada/nova-scotia
/donkin-mine-safety-violations-1.6718524.

CBC News, "Coal by the numbers," April 6, 2010, www.cbc.ca/news/canada
/coal-by-the-numbers-1.935272.

CBC News, "Springhill icon [Norma Ruddick] passes away," September 20,
2012, www.cbc.ca/news/canada/nova-scotia/springhill-icon-passes-away
-1.1198961.

CBC News, "Springhill mine disaster," 2011, www.cbc.ca/player/play
/2157768816.

CBC News, Springhill mine disaster song (as sung by the granddaughter of
Maurice Ruddick), www.cbc.ca/player/play/966275139853.

Cuthbertson, Richard, "Nova Construction hopes to open Junction Road coal
mine," September 9, 2015, www.cbc.ca/news/canada/nova-scotia/nova
-construction-hopes-to-open-junction-road-coal-mine-1.3219861.

Jessop, Alan, "Geothermal energy from old mines at Springhill, Nova Scotia,
Canada," www.geothermal-energy.org/pdf/IGAstandard/WGC/1995/1
-jessop.pdf.

Library and Archives Canada, www.library-archives.canada.ca/eng.

McInnes Public Inquiry report, www.mininghistory.ns.ca/index2.htm#spring.

Murphy, Dr. Noel, 1958 interview with Maurice Ruddick, www.cbc.ca/listen
/live-radio/1-27-information-morning-ns/clip/15900471-archival-recording
-maurice-ruddick-speaking-springhill-mining-disaster.

Nova Scotia Archives, https://archives.novascotia.ca.

Nova Scotia Museum of Industry, https://museumofindustry.novascotia.ca
/collections-research/coal-mining.

Patil, Anjuli, "Last man out after 1958 Springhill mine disaster dead at 95,"
September 9, 2018, www.cbc.ca/news/canada/nova-scotia/last-man-out
-after-springhill-mine-disaster-dead-at-95-1.4816384.

SaltWire Network, "Did Springhill miners see the bump coming?" October
22, 2018.

Springhill Heritage Group, Pat Crowe's Heritage Corner newspaper columns,
www.springhillheritage.ca/Springhill_Heritage_Group/Heritage_Corner.html.

Springhill Miners' Museum, www.facebook.com/SpringhillMinersMuseum.

World Coal Association, www.worldcoal.org.

INDEX

Note: Page numbers in italics indicate photographs and illustrations.
Page numbers followed by an "n" indicate notes.

Brine, Harold (*cont.*)
  physical appearance, 47–48, *258*
  post-Bump experiences, 122–123,
    128, 142–145, 147, 153, 155,
    156, 189–190, 199, 219,
    238–240, 242–245, 251–255,
    257
  relationship with Maurice Ruddick,
    70–71
Brine, Joan (née Cormier; Harold
  Brine's first wife), 34, 39, 40,
  41, 49, 102, 143, 190, 215, 233,
  239, *258, 286*
Brine, Murriel (née Campbell; Harold
  Brine's second wife), 312, 324,
  327
Brotman, Adolph E. (*Life* magazine
  illustrator), 270–271
Buckland, A.W.J. (*Toronto Telegram*
  editor), 264–265
Burden, Arnold, 74, 78, 102, 113,
  118, 132, 133, 179, 207, 285
  appearance, 76, *81, 82, 173*
  early life and career, 76, 80–81
  Ed Sullivan Show appearance,
    279–281
  marriage, 83
  military career, 82–83
  rescue efforts, 135, 137, 168–174,
    248–255, 272–276
  retirement life, 315, 317
  work as a miner, 80, 83–85,
    90–91, 97, 110
Burden, Helen (née Dewar), 76, 83,
  101, 273, 279, 315
Burden, William (Arnold Burden's
  son), 77
Burton, Charlie (mine overman), 11,
  107, 216

C

Calder, George (Springhill mine
  manager), 92, 115–117, 118,
  125, 133, 136
  rescue operations, 164–169,
    248–250, 252
Calder, John, 28, 275, 330n
Calder, Peter (George Calder's son),
  301, 309–310
Caldwell, Sam (Georgia state official),
  263–264, 291
Campbell, Bertha, 18
Canadian Legion Ladies' Auxiliary, 177
Canadian Press, 140, 234, 260, 287,
  296, 297
canaries, in coal mines, 43–44
Cape Breton, 3, 16, 17, 124, 181,
  298, 309, 316, 318
Carnegie Hero Fund Commission,
  306–307
Carter, Randolph, 135
Casey, Mary, 140, 336n
CBC (Canadian Broadcasting
  Corporation), 8, 75, 101, 114,
  182–184, 185, 225, 234, 247,
  253, 263, 287, 306, 308, 329n
CBHT (Halifax CBC television station),
  183
child labour in mines, 42
CHNS (Halifax radio station), 246
CKCW (Moncton CBC television
  station), 101
Clarke, Garnet (one of "the Seven"),
  111, 129, 160, 161, 197,
  200–202, 203, 221–222, 223,
  229–30, 237–238, 274, 305
coal, 15–16, 17, 99
  varieties, 21–22
Cobequid Hills, 13–16
Cochrane, Harry (*Chronicle-Herald*
  reporter), 336n